CON

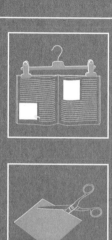

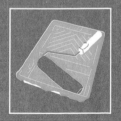

LIFE HACKS

Published in 2016 by
Harper Design
An Imprint of HarperCollins*Publishers*
195 Broadway
New York, NY 10007
Tel: (212) 207-7000
Fax: (855) 746-6023
harperdesign@harpercollins.com
www.hc.com

Distributed throughout the world by
HarperCollins Publishers
195 Broadway
New York, NY 10007

ISBN 978-0-06-240532-6
Library of Congress Control Number

Printed in the USA

First Printing, 2015

LIFE HACKS

Helpful Hints to Make Life Easier

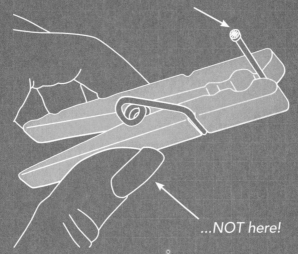

Bring hammer down here...hard

...NOT here!

Over 130 amazing hacks inside!

Dan Marshall

HARPER
DESIGN
An Imprint of HarperCollins Publishers

INTRODUCTION

Welcome to *Life Hacks*, the book for people who long to be free from those little worries and annoyances that complicate life on a daily basis.

How often have you wondered how you could chill your Chardonnay without watering it down with ice cubes, whether you actually locked your front door when you left, or who it was that borrowed your only copy of *The Goonies*?

The idea of "hacking" our lives, that is, finding clever ways to solve annoying problems and make mundane tasks easier, has been around since early humans started making fires and using simple tools. Ever since then, we have surrounded ourselves with time-and-effort-saving inventions and techniques, including my personal favorite: the remote control. Thank the stars that someone recognized the need for a device that does away with the effort of walking three feet to the television to change between channels. However, if, like me, you continually misplace one of these "magic boxes," fear not—there is a hack within these pages to ensure that this problem will never plague you again.

So read on to discover this and many more ingenious ways to make your life that little bit easier!*

*NB—these "hacks" are simply ways to make certain things a little easier, not revolutionize your entire life; you will still have to exercise, eat your greens, and wash occasionally.

HOUSEHOLD HACKS

Everyone has something in the home that could benefit from being hacked–from moving heavy objects to expanding your wardrobe space, there is something for everyone in this section, including a hack that will have women (and possibly a few men, too) digging out their shoes and blow-dryers. I won't go into detail, but suffice to say you won't be disappointed!

WRAPPING PAPER CLASPS

After my careful gift wrapping has been done I want to store my leftover paper (there's always some!) safely away for next time, without it ending up all unraveled and crumpled. Here's how...

Simply cut a spent toilet paper roll lengthways and wrap it around the roll of wrapping paper. This keeps the paper neatly in place and wrinkle free. Just try not to cringe when your friends and family carelessly rip the perfectly preserved paper to shreds to get to their precious gift.

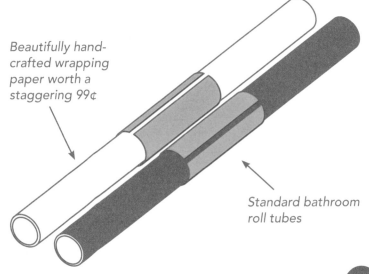

Beautifully hand-crafted wrapping paper worth a staggering 99¢

Standard bathroom roll tubes

RUBBER BAND LATCH CATCH

Doors are useful for getting into rooms. Except, that is, when they're broken. If you have experienced the anger and inconvenience caused by a faulty latch on a door (especially if it sticks and traps you in the bathroom) then you'll be pleased with this hack.

Loop a strong rubber band around the base of the handle on one side of the faulty door, twist it over, and then loop it around the base of the handle on the other side. The rubber band will make a latch-catching barrier where it forms a cross, which will hold your faulty latch back, avoiding any door-sticking scenarios. It's also handy if you're a parent and want to be able to burst into your teenager's room without warning.

Door handle

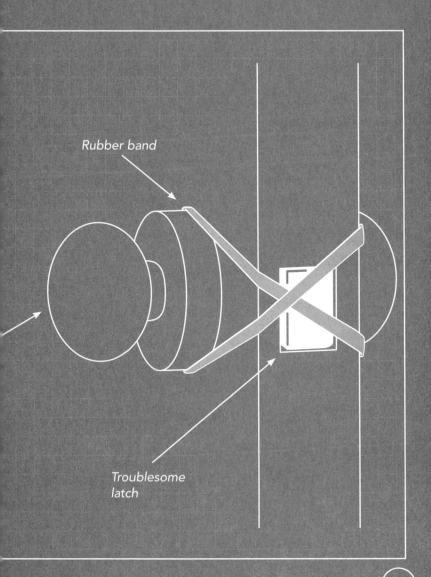

Rubber band

Troublesome latch

11

HOLIDAY LIGHT HANGER

Never again will you spend hours untangling the tree lights—this hack will prevent knots and makes it easier to store them away. Two for one—a holiday steal!

When packing away the lights, try wrapping them around a coat hanger. Not only will this keep them tangle free but will also give you the option of hanging them up somewhere, so you don't have to hunt for them next year.

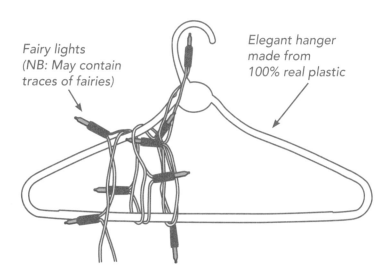

*Fairy lights
(NB: May contain
traces of fairies)*

*Elegant hanger
made from
100% real plastic*

HEAVY-LIFTING HELPER

Having moved several times, I must confess that this next hack never crossed my mind. I'm an idiot. An idiot with a bad back. Here's how it should be done.

If you're packing up to move or simply rearranging your excessive amount of worldly goods, load heavy items into a suitcase with wheels and pull it to wherever it needs to go. The last thing you need is to load up a heavy box only to have the bottom fall out on you. Heavy things hurt when they land on your toes.

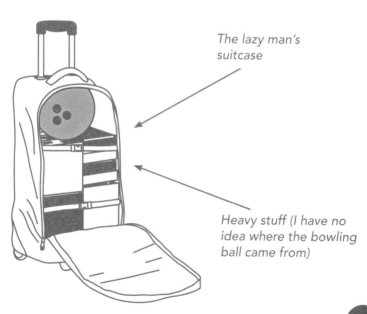

The lazy man's suitcase

Heavy stuff (I have no idea where the bowling ball came from)

SMALL SHOES, SMALL PROBLEM

This is one for the ladies, and possibly the men, too. Internet shopping has a lot to answer for. You've just bought some new shoes online, the last pair in your size, and lo and behold, they're just a little too small. Don't panic, help is here with this clever hack.

To get the shoes to fit, put on several pairs of socks and squeeze your feet into the overly tight shoes–this tip works especially well with leather but is also effective for other material types. Grab your blow-dryer and blast your feet with hot air for about ten minutes (be aware this may get a little hot, so use caution!). The stretching effect, combined with the heat, should help to loosen the shoes and make them fit better.

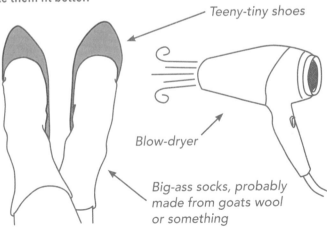

Teeny-tiny shoes

Blow-dryer

Big-ass socks, probably made from goats wool or something

KEY-RING CRACKER

Sometimes I feel the need to add a tasteless novelty key ring to my keys but encounter difficulties in prying the key ring apart. Anyone else?

To overcome this inconvenience, use a staple remover. Wedge the "teeth" of the remover in between the clasp of the ring, making it as easy as pie to slip on your new addition. Personally, I recommend a pewter Winnie the Pooh.

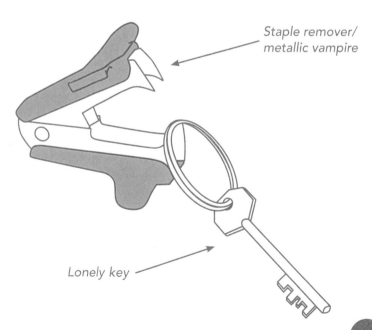

Staple remover/
metallic vampire

Lonely key

SMART CLOTHES STORAGE

I have T-shirts that haven't seen the light of day in decades, purely because I put clean clothes on top of the stuff already in the drawer. This organizational hack will prevent you from missing out on wearing that T-shirt that no one likes and help you to rediscover old favorites that have been hidden for years.

Arrange folded clothes within your drawers so that they are upright instead of layered on top of each other, so all items can be seen at a glance.

Awesomely discoverable clothes

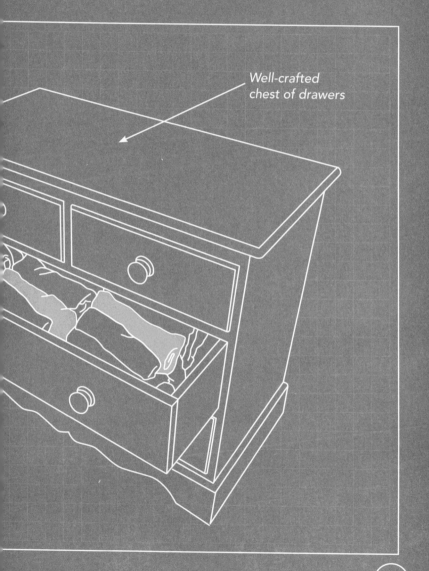

Well-crafted
chest of drawers

SMALL ITEM RETRIEVAL SYSTEM

Where do all those earring backs, watch screws, and contact lenses go when you drop them? I'll tell you where: They're still there, it's just that you're too blind to see them.

When you drop something small and can't find it, grab your vacuum cleaner and a pair of old tights. Slip the tights over the vacuum nozzle and fix in place with a rubber band. Run the vacuum over the area where you think you dropped your item and, with a bit of luck, the item will be sucked onto the tights where you can pick it off with ease. If you find anything of value that isn't yours then remember this: finders keepers, losers weepers.

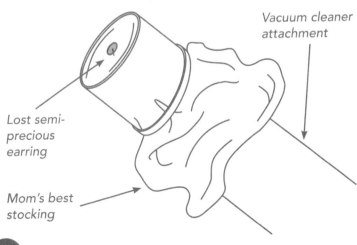

Vacuum cleaner attachment

Lost semi-precious earring

Mom's best stocking

COLLAR STRAIGHTENERS

Have you ever put on a shirt only to discover that the collar has creases in it? Don't waste precious minutes by breaking out the ironing board–fix it the *Life Hacks* way!

Grab a flattening iron (don't bother asking your sister/girlfriend/mom beforehand, they will only say no!), turn it on, allow it to reach full temperature, and then proceed to use it as a mini-iron by clasping your collar ends and "straightening" them out. And be careful: hot flattening iron + bare skin = bad times.

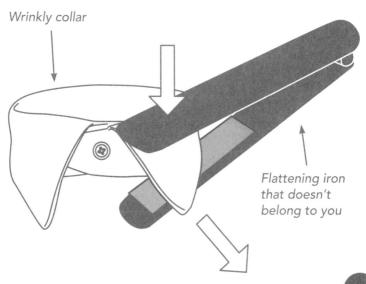

Wrinkly collar

Flattening iron that doesn't belong to you

SILLY SAFETY ROUTINE

When leaving the house for extended periods of time, I get nervous and can never remember if I've locked the door or not. This hack will save you returning home when you're halfway down the highway.

Vacate your premises as you normally would and lock the door. Now comes the fun bit: do something unusual to make a mental note that it is done. Try turning around three times or walking backwards down the path. That way, if you get the panicked urge to think back to whether you locked it or not, you will remember your silliness and thus feel at ease. I do the caterpillar down the path. I'm pretty sure the folks at No. 36 think I'm crazy.

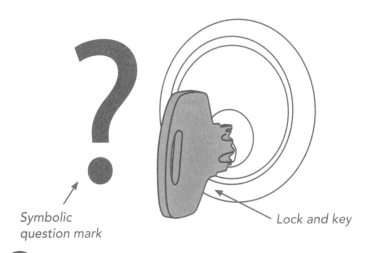

Symbolic
question mark

Lock and key

CHEWING GUM REMOVER

You know the scenario; you sit down on the bus or train only to discover that some little scallywag has left chewing gum on the seat. Your clothes are ruined. That's another trip to the charity shop you can't afford! But wait, there's a hack on the horizon...

Once you're home, remove the gummed-up item of clothing and put it in the freezer. Leave it in there for about an hour or until the chewing gum is rock hard, and then just pick it off. Alternatively, you could pop into Iceland, put some frozen peas on your bum, and wait for the gum to dry up.

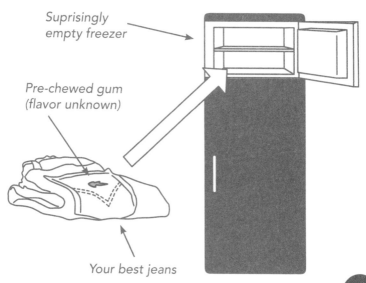

Suprisingly empty freezer

Pre-chewed gum (flavor unknown)

Your best jeans

MILK BOTTLE WATERING CAN

Fancy giving your impatiens a drink but can't find your watering can? The solution is only a skewer away.

Find an empty plastic milk carton. Remove the lid and pierce it with a metal skewer until you have sufficient holes to resemble the nozzle of a watering can. Fill the milk carton with water and attach the lid. You can now water the plants to your heart's desire or when the water runs out, whichever comes first.

Old milk container
(rinsed out)

Water

NOOK-AND-CRANNY VACUUM

Most of us need to clean out the almost inaccessible nooks and crannies of our electrical equipment that we cannot live without. This is how to do it without spending a fortune on compressed air canisters.

Attach the lid from a squeezy sauce bottle (the kind with a nozzle) to your vacuum cleaner. This will give you the ability to clear the crumbs from your keyboard or the dust from your phone's speakers with ease. I've always wondered what the difference between a "nook" and a "cranny" is–answers on a postcard, please!

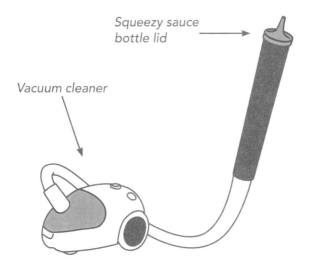

Squeezy sauce bottle lid

Vacuum cleaner

BEDTIME BUMPER

This hack will prevent many a bump in the night for anxious parents whose kids seem to involuntarily fall out of bed, having been transferred from their sturdy, four-sided cot.

Get hold of a pool noodle (that's the long cylindrical float thing used to aid swimming) and place it under a fitted sheet on the side of the mattress open to the room, creating a soft barrier. The noodle will prevent your beloved child from rolling over and out of the bed in the middle of the night, saving injury and adding bonus hours of sleep to your already starved routine.

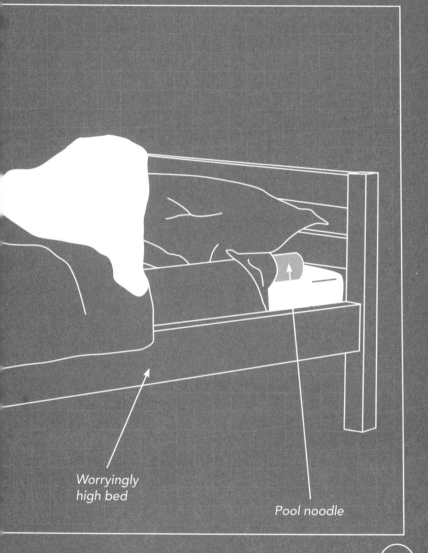

Worryingly
high bed

Pool noodle

WARDROBE EXPANSION KIT

I don't know anyone who hasn't suffered the stress and strain of not having enough closet space. Personally, I struggle every day with where to hang my freshly laundered mesh tank tops. At least, I did, until I found this hack.

Save the pull tabs from cans of soft drinks and beer and thread them onto the hook of a hanger, letting them rest at the bottom where the hook meets the hanger itself. You have now created a loop to attach another hanger, thus doubling the capacity.

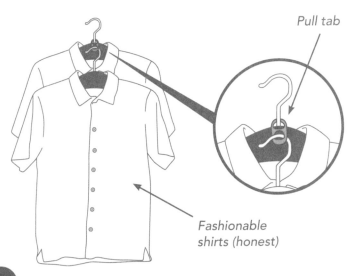

Pull tab

Fashionable shirts (honest)

SCISSORS SHARPENER

The scissors in my house are so blunt it's like I'm cutting with two lumps of wood. If you have the same problem, try this hack.

For the sharpest scissors in town, spend some time cutting shapes out of a folded-over piece of sandpaper, the coarser the better. The grain on the paper will act like a sharpening stone. Just remember not to run with your scissors once you're done.

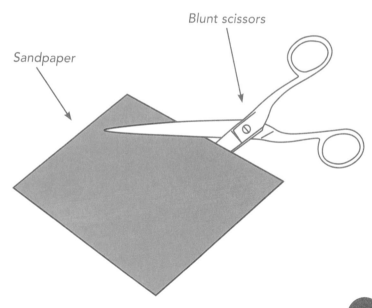

Blunt scissors

Sandpaper

CRAYON REMOVER

Crayon scribbles on the wall are the bane of any household with young children. But this way of removing them (the scribbles, not the kids) will put your mind at ease.

Take a cloth, spray a little water-displacing lubricant on it (the kind that comes in a bold blue and yellow can), and apply to the offending area. The crayon marks will magically disappear!

Water-displacing lubricant

Child's beautiful drawing of a fish

Crayons

VELCRO HOLDERS

Are you sick of losing the remote control? I have to say that sometimes I find mine in the bathroom. If only there was a way of making sure that it stays put…

Velcro is your friend for this hack. Attach some to the back of the remote (use the kind with an adhesive backing, available at all good hardware stores) and its corresponding part to the edge of the coffee table, or wherever you want to keep the remote (not the bathroom). And you don't have to stop at the remote–go wild and figure out what else needs a more permanent home!

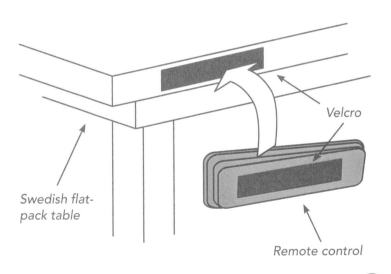

Velcro

Swedish flat-pack table

Remote control

CLOTHING DE-WRINKLER

Here's another effort-saving, iron-beating, clothes-related hack.

If the thought of ironing that wrinkled T-shirt is just too much to bear, throw it in the dryer along with a few ice cubes and set it to spin for five minutes. The wrinkles will magically disappear! Just make sure you don't try it with the "Grape Cubes" in the Food & Drink section (page 41), otherwise you'll end up with boiling hot grape goo on your clothes.

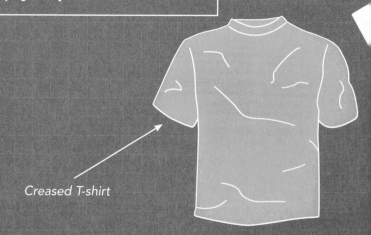

Creased T-shirt

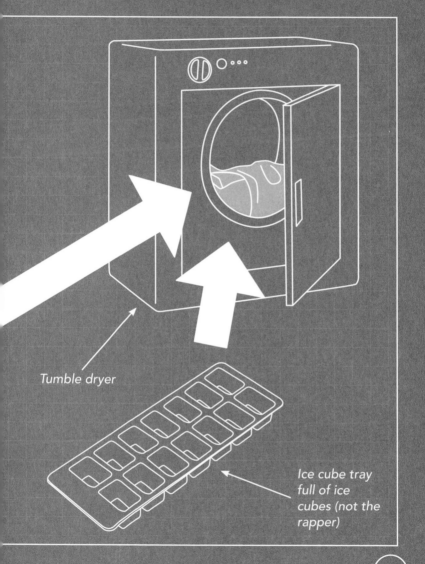

Tumble dryer

Ice cube tray full of ice cubes (not the rapper)

CAR DOOR BUMPER

Cars are expensive, especially when it comes to repairs. So, to save your hard-earned cash for more important things, like beer, why not reduce the likelihood of car damage?

Cut a pool noodle in half and stick it to the wall of your garage; that way, if you are a bit overzealous when opening the door once you've parked, you will prevent it from looking like it's been chewed by Jaws from James Bond.

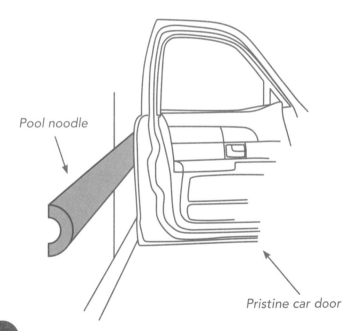

Pool noodle

Pristine car door

SPAGHETTI CANDLE LIGHTER

Have you ever burned your fingers trying to light a candle with a short wick and/or a tall candle receptacle, causing you to drop the match onto your Afghan rug, setting it on fire, and as a result you are now living under a bridge? No, me neither. But if you want to avoid all that pain and displacement, simply set fire to the end of a piece of spaghetti (uncooked, obviously) and light your short-wicked candle that sits in a tall receptacle (who makes those things?!).

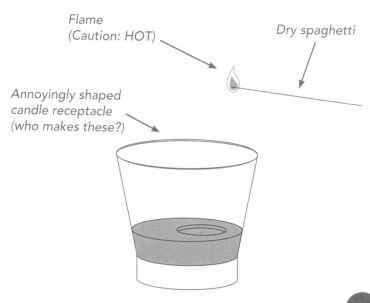

Flame
(Caution: HOT)

Dry spaghetti

Annoyingly shaped
candle receptacle
(who makes these?)

FOOD &
DRINK
HACKS

Who doesn't like food and drink? Personally, I can't live without them. This collection of hacks will show you how to make your fizzy drinks fizzier, your leftovers hotter, and your precious wine cooler. Feast your eyes on these top tips to make your scoffing superior and your slurping sublime!

STAY FIZZY

Like most people, I like my fizzy drinks to be, well, fizzy, and dread the idea of flat cola. The way to prolong your bubbly soda bliss, presuming you have opted for the plastic bottle version of your favorite drink, is to quite simply squeeze the air out of your bottle (once you've had a few delightful gulps), so that the liquid is near the top before replacing the lid. This gives the "fizz" gases nowhere to go and thus keeps your drink invigorated. Works for small or large bottles–not so much for glass ones.

Carbonated (that's posh for "fizzy") drink

LEFTOVERS–PIZZA PARTY

We all hate reheated pizza. The dough goes from crisp to chewy and you end up gnashing and tearing away at the crust like a lion at a gazelle's hindquarters. But worry not. There is a way to stop this unsightly way of eating.

When reheating your pizza, place a glass containing some water into the microwave with the pizza and the crust will not dry out. If you still want to tear away at it like a lion, that's up to you.

Microwave

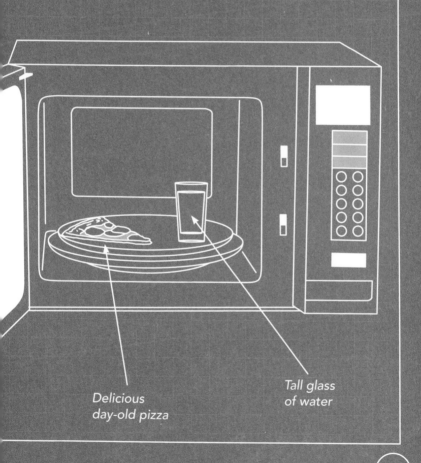

Delicious day-old pizza

Tall glass of water

BANANA-RAMA

We all love bananas, well some of us do. Wouldn't it be nice, then, if they lasted longer in the fruit bowl? You can't really munch your way through a whole bunch in a day–you'll get potassium poisoning! Let me put your fruit-loving mind at rest with this clever little hack.

By wrapping the stalks in plastic wrap the bananas will be fooled into thinking that they are back on the tree (this might not technically be true, as no one really knows what a banana is thinking). The bananas will now last three, or possibly four, extra days, giving your body the time to recuperate before you hit it with more bendy yellow goodness.

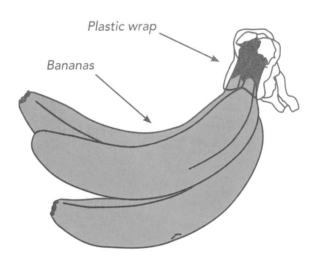

Plastic wrap

Bananas

BAKED "GRILLED CHEESE" SANDWICHES

If you're a cheese-on-toast fan or a *croque* connoisseur, this hack takes you one step beyond.

This version of a grilled cheese sandwich will not only make you salivate in anticipation but also toast the bread to perfection and keep the cheese all gooey and delicious. Prepare your sandwich as you would normally, using a cheese of your choice, wrap it in parchment paper, and place into a hot oven. Keep an eye on it to make sure it doesn't burn, and when it's done take it out and enjoy. Just be sure to blow on it first; melted cheese can be a bit "burny." The best part about this hack is that you can cook more than one sandwich at a time, so you'll never be short of grilled cheese. I'm eating one right now.

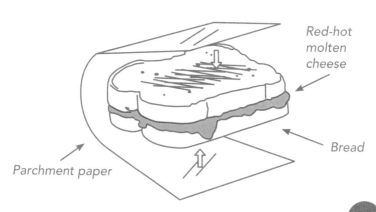

Red-hot molten cheese

Bread

Parchment paper

BANANA-THE EGG SUBSTITUTE

Here's a good one for the vegans among us—or those who simply must do some baking in the middle of the night, but have no eggs. Come on, we've all felt the urge to do some midnight cookie baking, right?

If the baking urge does hit you and the chicken coop is bare, you can simply substitute each egg with half a banana, which will act as the binding agent. This does tend to give whatever you're baking a hint of the curved yellow fruit, so if you're not a fan of bananas, this one's not for you.

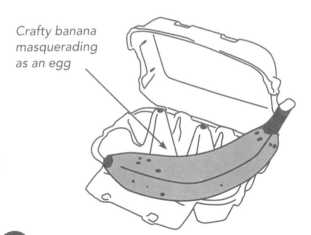

Crafty banana masquerading as an egg

GRAPE CUBES

Picture the scene: It's Friday night, you want to kick back with a nice chilled glass of Chardonnay (or fruit juice, for you teetotallers), but you forgot to put the bottle in the fridge to cool. D'oh! But you have the solution: grape cubes!

To chill your drink instantly, without watering it down to a tasteless mess with ice cubes, freeze some grapes and drop them into your glass. When they eventually thaw, they won't melt and water down your precious, precious alcohol.

Perfectly chilled wine

Frozen grapes

STEADY PEPPERS

So, you have dinner guests coming over and want to show off a little. Stuffed peppers always impress, right? But how do you cook them without making a complete mess of your oven when they invariably topple over and spill their delectable filling? Easy.

Dig out your muffin baking tray–its perfectly sized recesses will keep the peppers upright while cooking, as well as at the preparation stage. But be warned: If you are a bad cook, no matter how much you want to impress your dinner guests, this hack won't help. Do yourself a favor and order in.

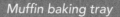

Muffin baking tray

Stuffed peppers

GOOD EGG, BAD EGG

Eggs are tricky little customers, perfectly packaged so we can't see or smell if they've gone bad until they're cracked open!

To avoid that "bad boiled egg smell" in the kitchen (and to avoid you or your dog being blamed for poor bowel control again), separate the bad from the good by placing the egg(s) into a bowl of water: The ones that float to the top are bad; the sinkers are good. I like to remember this information with a little rhyme: Floaters for the bin; sinkers for the win!

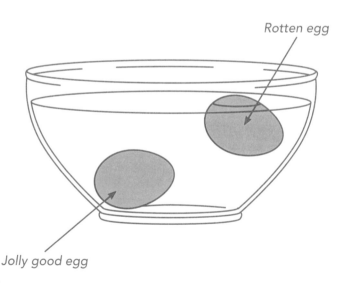

Rotten egg

Jolly good egg

PEELABLE EGGS

Keeping on the egg theme, this hack will make peeling hard-boiled eggs a cinch. No longer will you have to suffer the indignity of picking the eggshell off bit by little bit, constantly aware that at any moment you could fall foul to a piece of it jamming itself under your fingernail.

To avoid the frustration, not to mention the injury, try putting half a teaspoon of baking soda into the water while your eggs are cooking. This softens the shell slightly, making them miraculously easy to peel.

Tasty boiled egg

Easy-peel shell

LEFTOVERS–MICROWAVE MAGIC

How many of us have sat down in front of the television with our conveniently microwaved leftovers only to discover that somehow the food is cold in the middle, yet piping hot around the edge?

To avoid having to trudge back to the microwave to blast it again, try this hack out. Arrange the meal around the outside of the plate in a ring doughnut shape, leaving a hole in the middle. Now that there is no middle to stay cold, you will never again suffer a meal that is served at two vastly different temperatures.

Leftovers arranged in a pretty circle

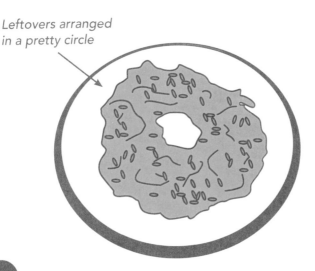

LONG LIVE THE KING (EDWARD POTATO)

Potatoes don't last like they did in my day…even the ones with fancy names. You can make them last longer with this hack.

To increase the shelf life of potatoes, just put an apple in the bag or container where you store them. The apple will not only provide some fruity company, but will also give off a gas that prevents the humble potato from sprouting. But when they've become shriveled and wrinkly like a giant tortoise's knee joint, then they are well and truly past it.

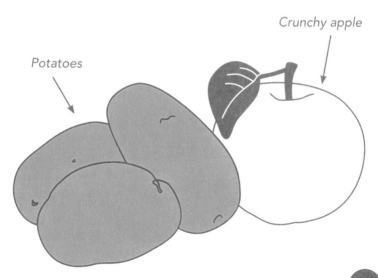

Crunchy apple

Potatoes

THE ULTIMATE TOASTED SANDWICH

If, like me, your love for the perfectly toasted sandwich knows no bounds, read on.

When toasting the bread, place two pieces side-by-side into one slot of the toaster and slide the handle down. What do you get? That's right: a half-bread, half-toast hybrid! Add your favorite filling (mine's bacon with ketchup, if you're wondering), and tuck in. This is quite the sandwich, giving you the best of both worlds—crunchy toast on the outside and soft doughy bread on the inside. Dee-lish!

Slices of "broast" →

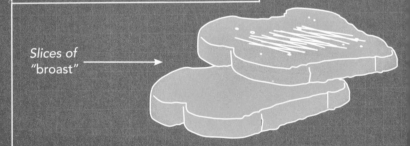

Slices of bread

Toaster

49

NO-MESS PANCAKES

Everyone knows that homemade pancakes can leave the kitchen looking as though Jackson Pollock has been creating his latest masterpiece in there. This hack will cut down on the mess and also leave you looking extremely cool in front of your friends and family.

Prepare your pancake batter and transfer it into an empty, cleaned-out squeezy bottle. When you want a fresh pancake, give the bottle a good shake and squeeze the mixture into the pan with no mess! A word of warning, though: Don't confuse the pancake bottle with the mayo bottle–the result will not be tasty.

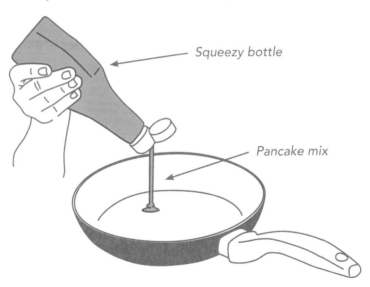

Squeezy bottle

Pancake mix

YOLK HOOVER

I have lost count of the number of times I have tried to separate the yolk from the white using the "eggshell" method, only to burst the yolk and ruin whatever it is I'm trying to make. This hack is much more effective.

Crack the egg into a bowl and get hold of a clean, empty plastic bottle. Remove the cap and give the bottle a squeeze, holding it tightly. While still squeezing, place the mouth of the bottle carefully on top of the yolk and release it. The vacuum you create will suck the yolk into the bottle to be transferred wherever you desire.

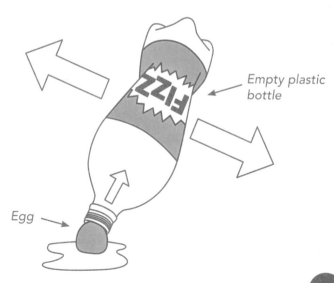

Empty plastic bottle

Egg

POPCORN TOOTH-SAVER

Do you love popcorn but hate it when you end up doing permanent dental damage on an un-popped kernel? This hack will help.

After popping the corn, don't immediately tear open the bag and eat it like a horse troughing from a nosebag. Instead, open the bag just enough for the unpopped kernels to come out, and give it a shake. You can then safely dispose of the tooth-hating, inedible kernels.

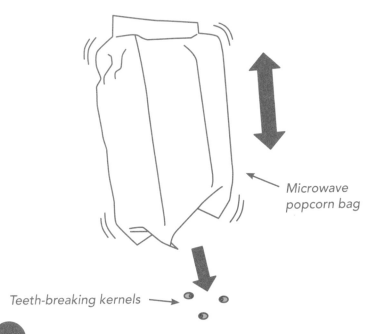

Microwave popcorn bag

Teeth-breaking kernels →

STRAWBERRY SKEWER

I have seen many an invention to hull or remove the stem from a strawberry, but none quite as ingenious as this. (The person who came up with those useless metal pinchers will be kicking themselves!)

To hull a strawberry (and I mean all of it, including the hard bit under the greenery), simply push a drinking straw from the tip of the strawberry all the way up to the stem. This completely removes the unwanted piece of fruit. Perhaps the clue was there all along... straw-berry.

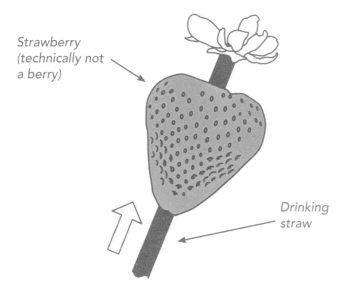

Strawberry
(technically not
a berry)

Drinking
straw

PAPER BEER-CHILLER

Your friends are coming over and the beer isn't cold–disaster! Don't worry, a simple paper-towel hack will save you from the wrath of those beer monsters.

Wet a paper towel and wrap it around your beer can/bottle. Place the beer in the freezer for fifteen minutes. When you take it out, the beer will be refreshingly crisp and cold for your guests–just remember to hide the imported beer before they arrive.

Freezer compartment

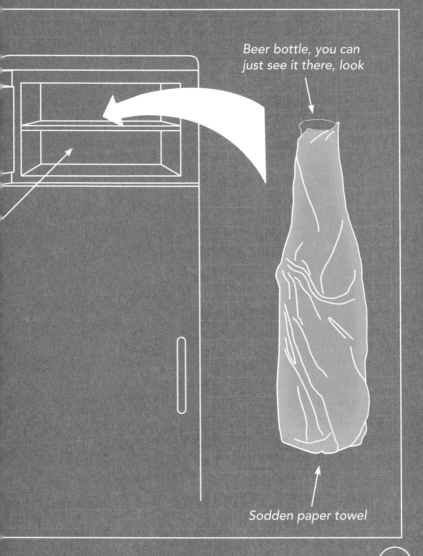

Beer bottle, you can just see it there, look

Sodden paper towel

ICE POP DRIP-CATCHER

Dripping ice pops are the bane of many people's lives during the summer months. The drips get everywhere–on your hands, clothes, steering wheel. Here's how to stop them.

A cupcake wrapper will help you survive this sticky predicament. Just poke the wooden stick down through the middle of the cake case to create a little cup to catch the offending drips.

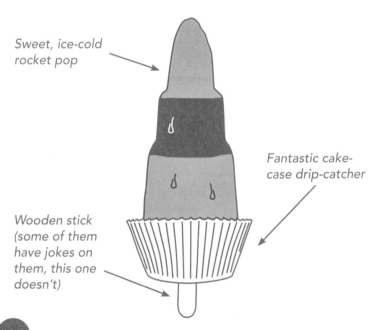

Sweet, ice-cold rocket pop

Fantastic cake-case drip-catcher

Wooden stick (some of them have jokes on them, this one doesn't)

SOFT DRINK STRAW-HOLDER

I hate it when my straw bobs up out of my drinks can and splashes sugary goodness all over my nice clean clothes! This ingenious idea puts an end to all that.

Simply turn the pull tab 180 degrees and poke your straw through the hole, thus holding it in place–this trick has saved me many, many embarrassing trouser stains. So simple you should be ashamed that you didn't think of it yourself!

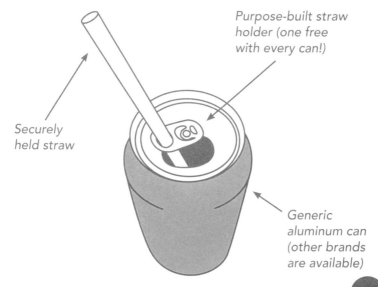

Purpose-built straw holder (one free with every can!)

Securely held straw

Generic aluminum can (other brands are available)

KITCHEN HACKS

Everyone needs help in the kitchen, even Gordon Ramsay. Not that you would say that to his face. But these hacks are designed to make even the mighty Gordon stop swearing and take notice. And perhaps make some of you swear in amazement!

BAGEL PROTECTOR

My bagel dreams are always shattered by lunchtime as I pull out my flattened excuse for a midday snack. Lucky for me I can now protect my bagel in something that looks like it was made for the job–except it wasn't!

When you have finished using all those CD-Rs, what are you going to do with the case? Or should I say bagel holder? That's right! Your bagel will fit perfectly in it, even utilizing the spindle in the middle. Perfect.

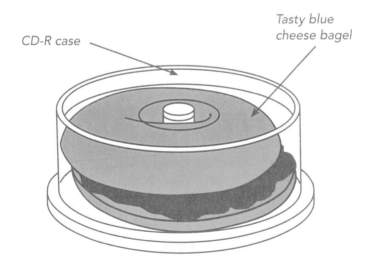

Tasty blue cheese bagel

CD-R case

FLOSS CUTTER

If you can't cut a straight line with a straight knife then you have no business holding one. Luckily there are other ways to obtain a clean cut when dividing soft food. Give the knife to a responsible adult and go to the bathroom. No, not to take a leak. Just grab the dental floss and head back to the kitchen.

Get a length of dental floss and use it like cheese wire to cut your cheese, cake, banana, etc. OK, there is a possibility that your food will taste a bit minty but at least it will be perfectly divided.

Dental floss receptacle →

How to hold the
dental floss

Perfectly sliced cake

REPURPOSED BAG CLIPS

Stale snacks are nobody's friend; shoving a handful of chips or nuts into your mouth only to discover that deliciousness has been swapped for foul chewiness is something no one relishes. Bag clips are available in shops to stop this happening, but why would you buy them when you can have them for free?!

Slide the clips off an old trouser hanger and use them to keep staleness at bay. You can now open a fresh bag, safe in the knowledge that you don't have to eat them all in one sitting–unless you really want to.

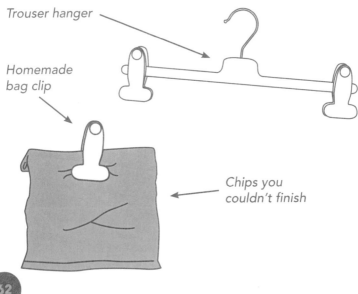

Trouser hanger

Homemade bag clip

Chips you couldn't finish

NONSLIP CUTTING BOARD

Chopping carrots in the kitchen is good–chopping fingers is bad. You can minimize the risk of this happening with something as simple as a paper towel.

Wet a paper towel and place it under your cutting board; this will stop the board from slipping and save your precious digits. Vegetarians will thank you for not turning them into cannibals.

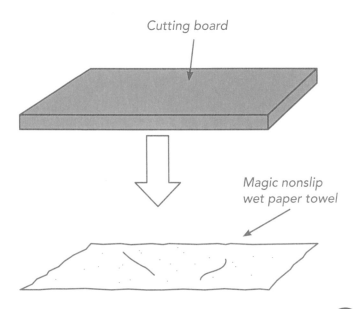

Cutting board

Magic nonslip wet paper towel

SUPER STORAGE RACK

Keeping cleaning supplies in order is paramount for any household. The shelf I keep my products on has stuff at the back from 1998!

To ensure you've got what you need when you need it, use a fabric hanging shoe rack (the kind that attaches to the back of a door) to store your products in. Ideally, place it on the inside of a high cupboard, so your cleaning products are uber-organized, accessible, and out of reach of little children.

Shoe rack (not to be confused with tie rack or Iraq)

Cleaning products

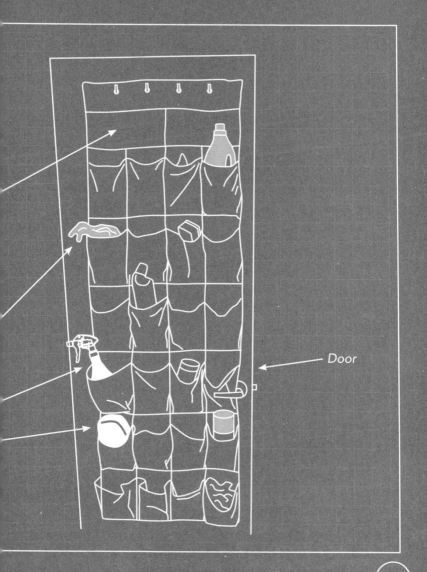

Door

EASY-EMPTY GARBAGE CAN

At home it's my job to empty the bin, and boy do I hate it! When you have stuffed as much in there as humanly possible, trying to get the bag out can be tricky. This is due to the vacuum that is created when trying to yank the bin liner out. To rectify this little problem get your power drill out.

Drill a couple of holes in the bottom of the bin to stop the vacuum ever forming. The garbage bag will now lift out with ease.

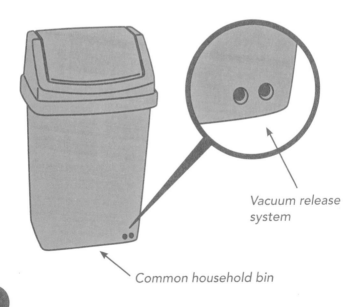

Vacuum release
system

Common household bin

COOKBOOK HOLDER

Are you annoyed at cookbooks that won't stay open? Not to mention cluttering up your prep area. Most of the pages in my cookbooks are completely stuck together–gross.

To keep books pristine, use a trouser hanger with clasps and attach to the tops of the pages to hold your place–then hang it on an overhead cupboard door to keep it off the surfaces.

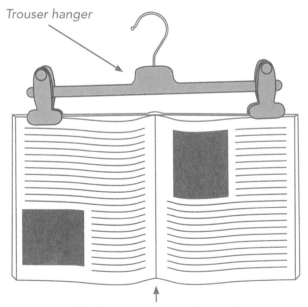

Trouser hanger

Third-generation recipe book

PHOTO SHOPPING

I've seen many a clueless shopper standing in the aisle of the local supermarket, scratching their head trying to figure out whether they need that bag of peas or not. These people are dumb, but I bet they have a smartphone in their pocket (see what I did there?).

To save yourself time, take a photo of the contents of your fridge and cupboards on your phone before leaving to do the weekly shop. That way, instead of scratching your head aimlessly, you can check the photos to see what you need and be on your way.

Refrigerator

Stuff you
don't need
to buy

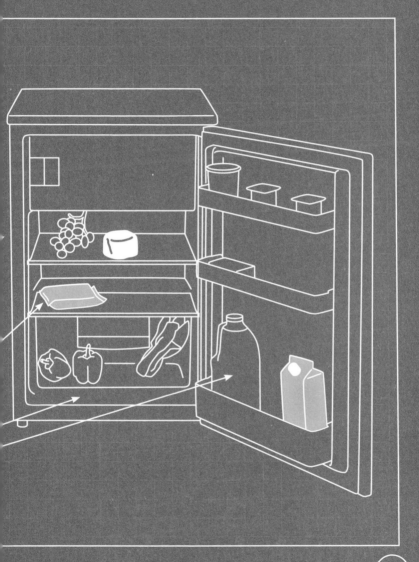

QUICK CONDIMENT TRAY

If you're planning a barbecue or garden party, spare a moment to consider the condiments–a mess of sauce bottles is unsightly, and do you really have enough ramekins to cover all the bases? Well, why not use a ready-made condiment tray?

Grab your muffin tray and simply serve ketchup, mustard, onions, etc. in there. Although, you will be limited to six items–unless you have another muffin tray stashed away, you sly old dog you!

Muffin tray

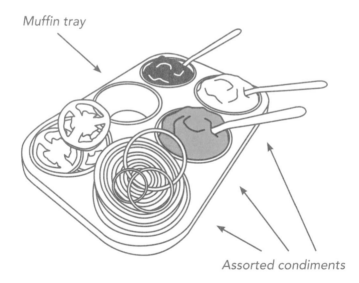

Assorted condiments

BOTTLE-TOP FOOD PRESERVER

When you need to keep your opened packet of foodstuff airtight, this hack is invaluable. Ideal for sealing bags of snacks, packets of pasta–or, in my case, king-sized bags of toffees.

Cut off the top third of a small plastic bottle to create a "collar." With the lid removed, push the top of the open packet through the neck of the collar. (Pay attention, here comes the good bit.) Fold the end of the packet back over the edge of the neck and replace the lid to create an airtight seal. Genius or what?!

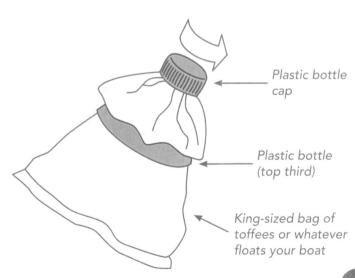

Plastic bottle cap

Plastic bottle (top third)

King-sized bag of toffees or whatever floats your boat

SIX-PACK DINING PACK

When dining al fresco (or, in my case, sitting on tiny chairs in a cramped garage) make it feel like you're eating out in a country pub garden–and save yourself trips back and forth to the kitchen–by using a makeshift cutlery caddy.

To create your caddy, simply save a cardboard six-pack holder (that's beer, for all you teetotallers). Put napkins, cutlery, salt and pepper shakers, and sauce bottles in the compartments and there you have it, authentic pub dining at home. Mine's a pint of Snaggletooth!

Accompaniments —

The trusty six-pack holder —

Utensils

BEER

BOILING OVER

Cleaning burned water off a cooktop is not a job I relish. To be honest, I'm pretty much anti-cleaning across the board. Luckily for me–and you–I have a host of hacks to save the trouble, including this one.

To stop the water from boiling over and messing up the cooktop, put a wooden spoon across the middle of the pot. This bursts the bubbles coming up that are trying to escape, stopping the liquid in its tracks.

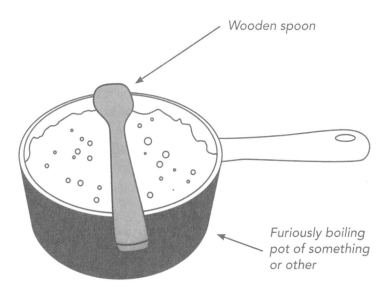

Wooden spoon

Furiously boiling pot of something or other

DIY
HACKS

Personally, I can't stand the thought of DIY. All that preparing, measuring, choosing the right tool–and that's before you've even started. And you're bound to put a hole in something you didn't mean to. I'd much rather be doing something interesting, like rearranging my sock drawer. But what's this…*Life Hacks* is here to save the day! (Unfortunately, you still have to do most of the work yourself, unless you have some sort of DIY-loving robot army.)

BLISTER PACK CRACK

Blister packs have got to be the hardest thing to get into, bar none. Even giant scissors can fail at this task. But there is a way.

Take your blister pack and use a can opener along the side of it, thus opening the outer edge of the plastic monstrosity with comparative ease. Take that, evil blister-pack inventor, you've been hacked!

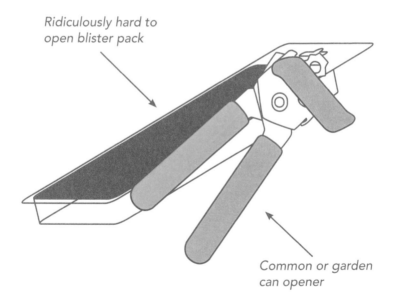

Ridiculously hard to open blister pack

Common or garden can opener

CORD TIE-IN

It is infuriating when you are stretching to drill in that hard-to-reach place and the drill loses power as it disconnects from its extension cord–it's also a health and safety no-no. This easy hack will ensure this never happens again.

Simply tie the two cords together, one cord over the other, at the connector end, before plugging it in to the wall. The more you pull, the more they stay together, and the power stays on.

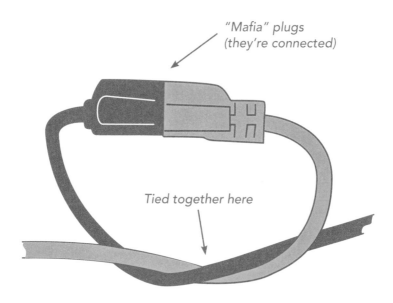

"Mafia" plugs
(they're connected)

Tied together here

THUMBSAVER–PART ONE

How to solve the age-old problem of hitting your thumb while hammering. Easy peasy–in fact, so easy that there are two solutions.

To avoid blackening your thumbs, use a clothespin to hold the nail in place before you strike it. Then you can thrash away with your hammer until the nail is firmly in place.

CAUTION: This hack will not help you if your aim is so bad that it bends the nail, in which case use a screw or buy a nail gun. On second thought, don't buy a nail gun…

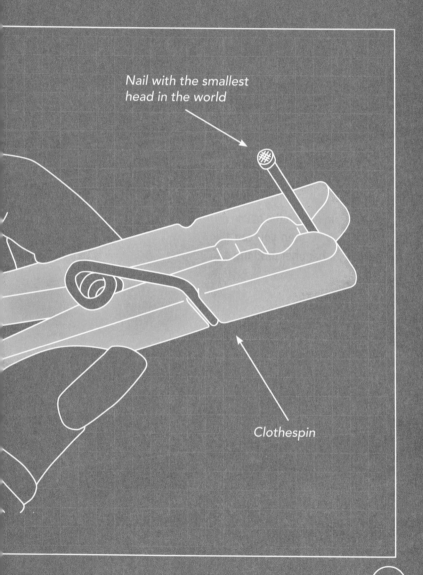

Nail with the smallest head in the world

Clothespin

THUMBSAVER-PART TWO

The second way of solving this problem is for those without access to a clothespin (how do you people dry your clothes?!). Simply take a comb and hold the nail in place by wedging it between two of the teeth. Don't tell me you don't comb your hair either?

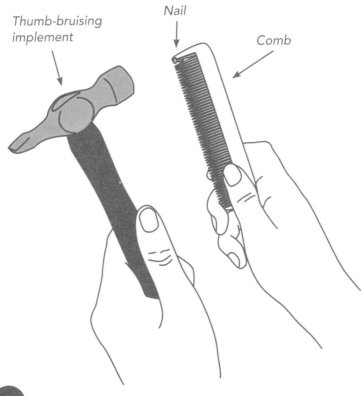

Thumb-bruising implement

Nail

Comb

MAGNET NAIL FINDER

I had a friend who once tried to hammer a nail on top of another nail; the hammer recoiled so violently that it hit him in the head, knocking him unconscious. Talk about Hammer Time!

He could have saved us a trip to emergency room by using a magnet to locate the hidden nails and studs already in the wall. If you hover over an area and the magnet sticks, you've got a nail!

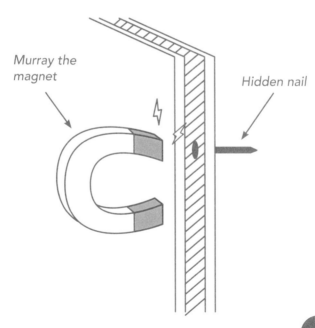

Murray the magnet

Hidden nail

NO-DUST DRILLING

The dust from drilling holes in walls can get everywhere. I've even found some on the cat. Here's how to make drilling dust free.

When the drilling urge comes over you, grab your trusty drill and a pack of sticky notes. Mark where you want to drill, stick a sticky note underneath, and then fold the note up to form a little trap for the dust. When you've finished drilling the hole, empty the dust in a bin, and reattach the sticky note beneath the next drilling point. This is a hack that keeps on giving!

Sticky note

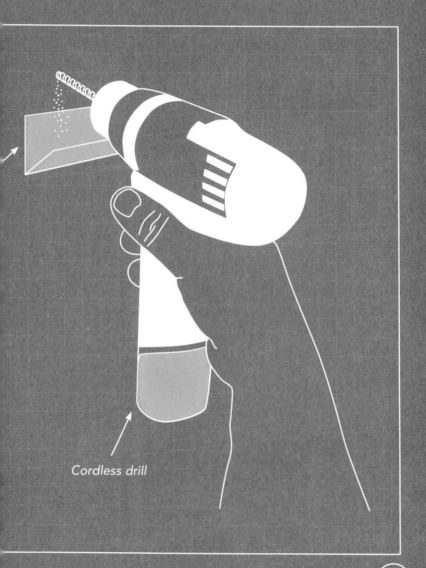

Cordless drill

PICTURE HOOKED

No picture hook? No problem. Here's another pull-tab hack.

Use the pull tab as a makeshift picture hook by securely screwing it to the back of whatever you're trying to hang. Chances are this will be more durable than those rubbish brass picture hooks, so I'm going to put that down as two wins.

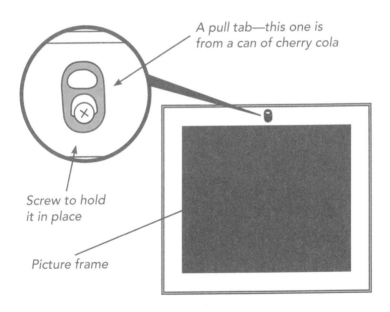

A pull tab—this one is from a can of cherry cola

Screw to hold it in place

Picture frame

EASY-CLEAN PAINT TRAY

Paint trays are ruined after a few uses—old paint builds up and those useful ridges just disappear. Here's how to keep your tray clean and functional.

Line the tray with tinfoil before pouring in the paint. This will protect the tray itself and still allow the grooves to do their work. When you're done, simply remove the foil and throw it in the bin. Your paint tray will remain like new for years—unlike the walls, which will need a second coat, possibly a third. Better buy that jumbo roll of tinfoil.

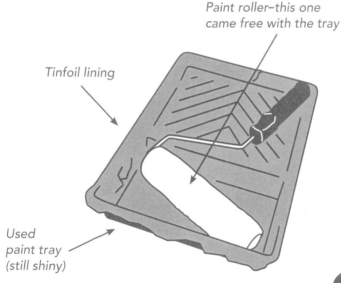

Paint roller—this one came free with the tray

Tinfoil lining

Used paint tray (still shiny)

ANTI-DRIP PAINT CAN

Fans of painting and decorating will like this one. Wiping the excess paint off your brush onto the side of the can might seem like a good idea at the time, but this results in paint buildup, which is messy and makes the lid stick when you try to replace it.

To solve this, place a rubber band across the opening of the can. This creates a handy scraper for wiping the excess paint off your brush. When you have finished painting, simply remove the rubber band and fit the lid back on with ease.

Paint brush

Dirty rubber band, dripping with paint

Immaculate paint can

HEADLIGHT BRIGHTENER

Fogged-up headlights on a car you're trying to sell are a nightmare–they look bad and show the car's age. (With my clapped-out piece of crap, this is the least of my worries, but I like to make improvements where I can.)

To have those headlights looking like new, grab yourself some toothpaste. Yes, that's right, toothpaste. Scrub the headlights with a small amount, rinse, and watch them sparkle.

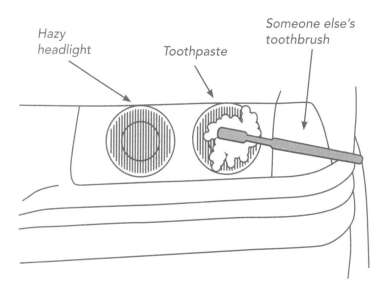

Hazy headlight

Toothpaste

Someone else's toothbrush

WALNUT WOOD-WAXER

If you have tired, scuffed wooden furniture at home then get ready for an all-natural hack that will blow your mind!

Rub a walnut kernel over the marks on the wood and watch them disappear before your very eyes. Obviously, if your furniture has been chewed to death by your dog, or scratched to pieces by your cat, then the walnut is useless. But you will have a tasty, healthy snack!

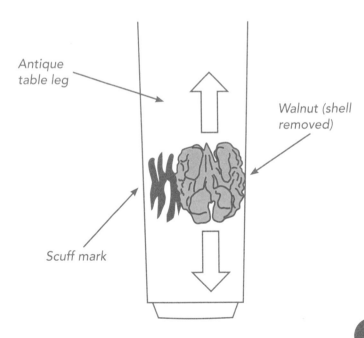

Antique table leg

Walnut (shell removed)

Scuff mark

BATHROOM HACKS

Is your bathroom a spa-like sanctuary or a germy embarrassment? Either way, the following hacks will help you upgrade its usefulness. Unfortunately, there's no hack to eradicate that eye-watering pong after a particularly long toilet session—for that, you may need to take industrial measures.

CAT-PROOF TOILET PAPER

Aah cats, they're so cute…except when you return home to find that they've left "a gift" in the hallway–then, when you run to get some toilet paper to dispose of the gift, you discover that the cat has also unraveled the entire roll!

This hack won't help prevent your cat from gifting dead animals, but it will help to keep the toilet paper intact. If you tuck the end of the roll into the cardboard tube, it will tuck away the flap that the cat loves to play with, and so hopefully stop it unraveling.

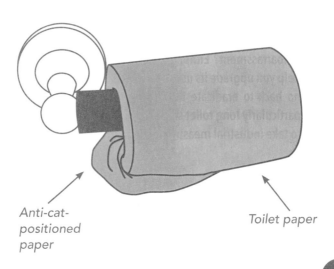

Anti-cat-
positioned
paper

Toilet paper

HOMEMADE CLEANER

Bathroom cleaning products can be expensive, so why fork out when you can use everyday household items to create your own. "What?!" I hear you cry. It's true…

Fill an old spray bottle with two-thirds vinegar and one-third washing-up liquid and you have just made your own limescale-beating bathroom cleaner. Unfortunately, you will still have to use a considerable amount of elbow grease to scrub the scum from your bathtub.

SUDS

Washing-up liquid

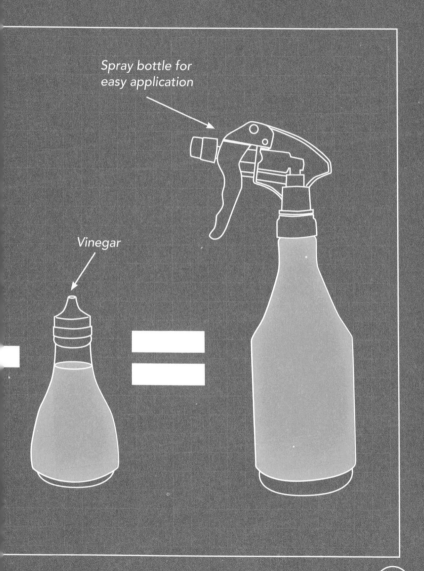

Spray bottle for easy application

Vinegar

RECYCLED SOAP

I have seen my grandfather use this hack for years. That thin bit at the end of a bar of soap, which looks like it will disintegrate as soon as you touch it, really can stretch further–here's how.

Simply take the sliver of soap and apply it to a fresh bar, using a little water to make it stick. You can now enjoy the satisfaction that you've scrimped and got one over on the soap company.

New soap

Crusty old soap

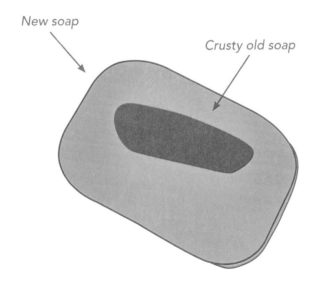

EAU NATUREL

This hack is for the naturist in all of us—and will leave your bathroom smelling clean and fresh.

Help yourself to a few branches of eucalyptus and hang them from the showerhead. As you shower, the warmth and moisture will disperse the fragrance from the leaves, giving you a natural, sweet-smelling boost to your showering. For all those smarty-pants who just said "Eucalyptus trees only grow in Australia!"—this is utter rubbish. The Internet says so.

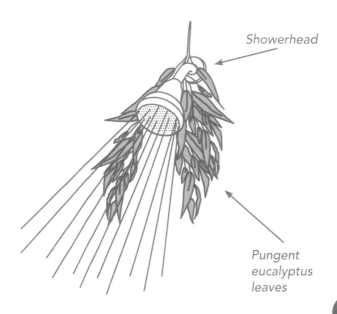

Showerhead

Pungent eucalyptus leaves

BATHROOM BUCKET FILL

When it's time to break out the mop and bucket in the bathroom, you'll have enough hassle getting into all those gross nooks and crannies–the last thing you want is extra effort lugging a scalding-hot bucket of water from the kitchen. Here's how to avoid it.

Place a bucket on the bathroom floor and put a dustpan in the sink (for this to work it has to be the type of dustpan where the brush can fit into the handle). Place the edge of the pan under the tap(s) and position the pan handle so it's jutting out over the edge of the sink. Now turn on the taps. The pretty waterfall it creates will make for a good temporary distraction while the bucket fills up.

Bucket
(don't kick it!)

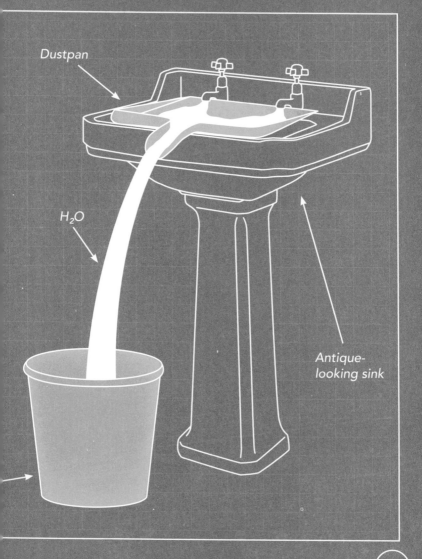

Dustpan

H_2O

Antique-
looking sink

VODKA... FOR HAIR

Many people like a drink, but washing your hair with booze is a bit over the top, even for the most radical enthusiast. However, doing this actually has health benefits, as you will see.

Add a shot or two of vodka to your shampoo and it will not only strengthen your hair but also stop your scalp drying out and help cure dandruff. Just don't be tempted to drink the concoction–fruit-scented shampoo does not make a good mixer.

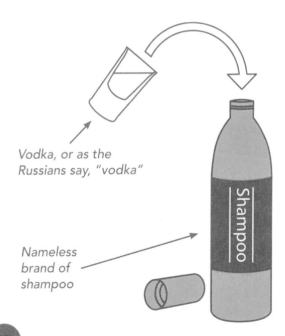

Vodka, or as the Russians say, "vodka"

Nameless brand of shampoo

Shampoo

SHOWER SMARTS

Become a little smarter while you shower with this little hack. (It doesn't involve waterproofing your textbooks or anything like that.)

Go out and buy a shower curtain with the map of the world on it so you can scrub up on your geography as you scrub yourself! Just don't try to take it out with you on your travels–it's simply not practical.

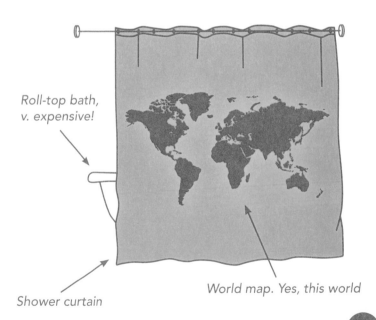

Roll-top bath, v. expensive!

World map. Yes, this world

Shower curtain

MIRACLE SHOWER-HEAD CLEANER

To stop your showerhead looking like something from the Bates Motel, try this next hack.

Fill a small plastic bag with vinegar, pull it over the showerhead, fix it in place with a rubber band, and leave overnight. The acidity of the vinegar will dissolve the scum that has built up and leave the showerhead looking like new. Remember to run the shower after you detach the vinegar bag–you don't want to end up smelling like a bag of fish and chips!

Plastic bag

Vinegar

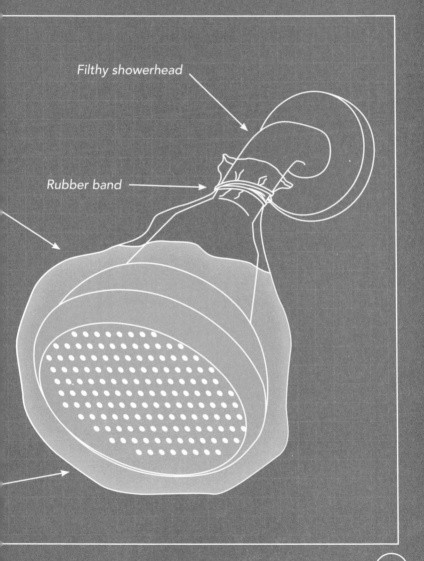

Filthy showerhead

Rubber band

FOG-FREE MIRROR

You've just had a nice relaxing hot shower and you're ready to get brushing/combing/flossing in front of the mirror. But, alas, you are going to have to do it "blind" as the mirror has now fogged up and you can't see jack. Sure you could wipe it off, but it's never quite right is it? You should have used this hack earlier.

Before you start your shower, try "cleaning" the mirror with shaving foam. Cover the reflective glass with the stuff and then wipe it all off again. Now when you step out of the shower, the mirror will be relatively fog free.

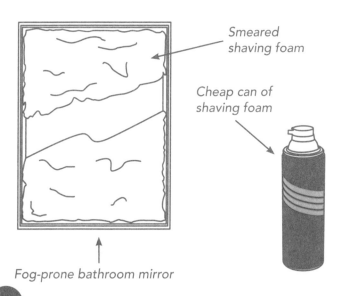

Smeared shaving foam

Cheap can of shaving foam

Fog-prone bathroom mirror

OFFICE HACKS

These hacks are for those of us who need some guidance in the office environment. Unfortunately, they won't help you rise through the ranks or catapult you into your ideal job–if career development is what you want, I suggest you get off your ass, slap your boss in the face, and tell him you're going off to follow your dream of opening a shrimp shack in Guatemala. But if that's too much for you, then read on to learn the handiest tricks and tips to make your working life a little better.

PASSWORD PROTECTED

Here's a hack to avoid hacking! Those pesky criminals have hacked your password…again. Well, it's unsurprising as you tend to stick to the most obvious ones: "password," "123456," "qwerty," or "monkey." If your password is actually one of these then you deserve to be hacked; in fact, would you like to win the Nigerian lottery? All you need to do is send your bank details to…

Seriously though, if you want to avoid being hacked just use an accented letter–â, é, ĩ, ö, ū, etc.–in your password. If the criminals can guess that then you're screwed, anyway.

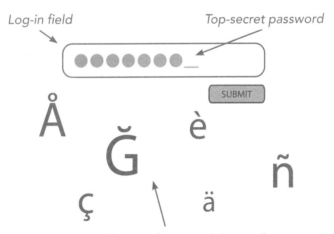

Log-in field

Top-secret password

SUBMIT

Å   Ğ   è   ñ   ç   ä

Accented letters, but you didn't really need me to point that out did you?

ABORT! ABORT!

Have you ever accidentally shut down your computer when working on important files that you've neglected to save? The IT gurus have created something that can help. Well, they've got to earn their pay somehow! (NB: This will only work if you're using Bill Gates's most famous operating system.)

Open up the notepad program and type in the following:

shutdown -a
pause

Save as a .bat file on your desktop. The next time you act like an untrained monkey and accidentally shut down the computer, click this file, and the shutdown will be aborted. This will save you precious minutes.

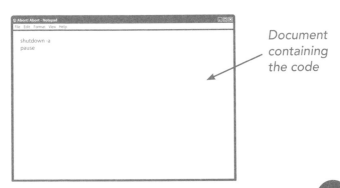

Document containing the code

KEYBOARD FEET

No, keyboards do not have actual feet. Who would make trainers that small?! I'm talking about the little foldout bits of plastic that angle the keys for better typing.

If one of these feet breaks, you'll have a wobbly keyboard–not ideal. To fix this, get hold of an aptly sized binder clip handle (the silver bit, minus the clasp) and wedge the two bare ends into the hinge holes where the plastic foot should sit. Problem solved.

Now, if only you could stop yourself hitting the keyboard with your face in frustration, then we wouldn't be fixing it in the first place.

Binder clip

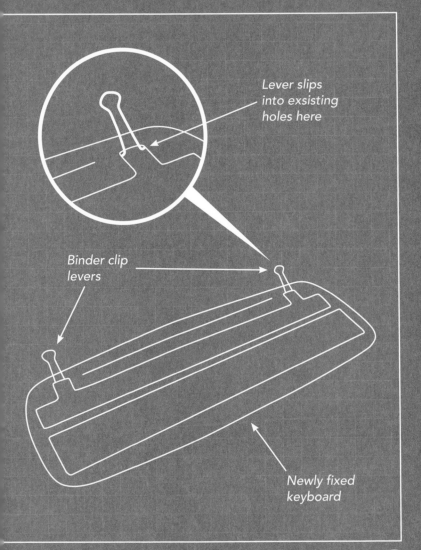

Lever slips into exsisting holes here

Binder clip levers

Newly fixed keyboard

CABLE CLASPS

Another genius use for binder clips is to sort out all those computer cables that you are constantly interchanging, like a switchboard operator from the 1950s.

If you attach the clips to the far side of your desk, you can feed your favorite wires through them so they are easily accessible. This will save you from having to crawl around on your hands and knees under your desk searching for the elusive buggers, especially inconvenient when your boss is around and/or your trousers are riding low.

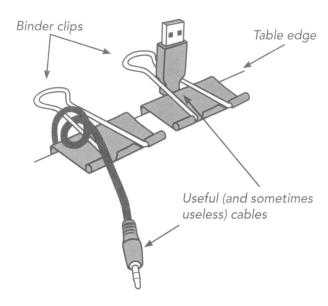

Binder clips

Table edge

Useful (and sometimes useless) cables

NO-MORE-KNOTS CABLE CADDY

Ever experienced "cable rage" as you try to untangle various connector leads before a big presentation in front of important clients? I've always wondered just how those cables manage to magically tie themselves into impossible knots!

To avoid anger and embarrassment, sensibly coil up your wires and keep them in an old glasses case or pencil case. This will stop the cables from moving around in your bag and getting all tangled together. You will also impress your prospective clients with your ultra-smart organizational skills!

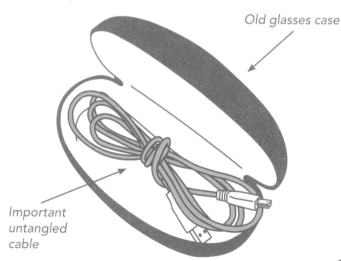

Old glasses case

Important untangled cable

IMPORTANT DOCUMENT PROTECTOR (IDP)

I, for one, don't like it when I'm asked to sign a creased, rain-soaked contract. One way to avoid this scenario is with this hack, the IDP.

Get a ziplock bag big enough to house your document and a piece of sturdy cardboard that's the same size or larger than the bag. Cut the piece of cardboard down to size if necessary, then place the cardboard inside the bag. You can now store your documents inside and keep them dry and wrinkle free.

You could even personalize your piece of cardboard, though I would keep it fairly corporate; Spongebob Squarepants doesn't look too professional.

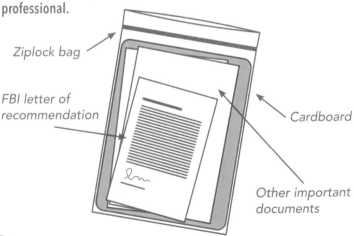

Ziplock bag

FBI letter of recommendation

Cardboard

Other important documents

THE SILENCER

If you're unfortunate enough to have an intolerably inconsiderate colleague–the kind who refuses to turn the volume down on their computer to an acceptable level–then before you head to HR, try this devilish hack.

Take a broken pair of headphones and cut the wires off at the jack–everyone who's ever owned a pair of earphones has a broken set, it's inevitable! Insert the now-wireless jack into the audio port at the back of their computer, then sit back and watch as they try to fathom why the sound isn't working! Try not to laugh maniacally, though, or the game will be up.

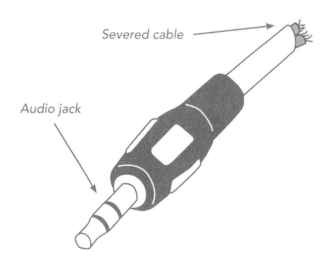

Severed cable

Audio jack

CABLE ORGANIZER

"More cables?!" you cry. Well, they are an integral part of office life, so here's another wire-related hack. IT technicians, especially, will benefit from this one.

Find yourself a sturdy box (the size should vary according to the amount of wires you have collected over the years) and begin collecting spent toilet paper rolls (enough to fill the box when standing them up vertically). When your collection is complete, fold your wires so that they fit snugly into the toilet paper rolls and place them in the box. You're on a roll!

Shoe box containing yet more toilet paper rolls and even more cables (all neatly packed away)

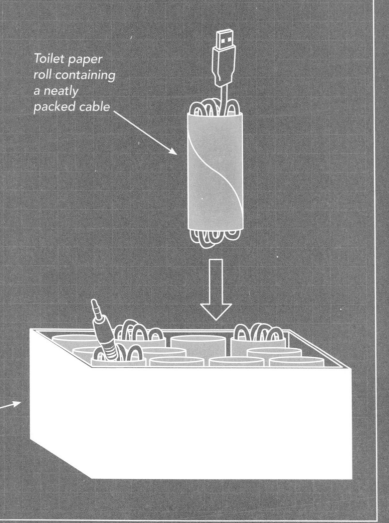

Toilet paper roll containing a neatly packed cable

PENCIL HOLDERS

This one is for freelance artists and school teachers–technically not office jobs, but the principle still applies.

To keep those coloring pencils organized, and avoid invoking the wrath of the "artistic temperament" or a full-blown classroom tantrum, store them in converted milk containers. Wash out the containers, cut out the lower quarters, and attach to the wall with a screw or a strong adhesive.

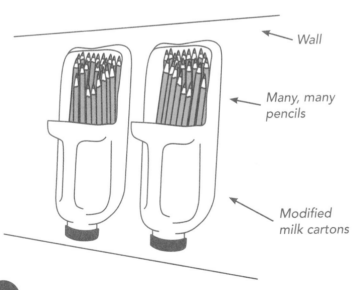

Wall

Many, many pencils

Modified milk cartons

COLOR-CODED KEYS

Just the other day I tried to get into my office using the key to my drinks cabinet, and not for the first time either! This hack will help separate the keys in your pocket and leave them looking colorful to boot.

Dig out the old nail polish (if you don't have any, go to a market where you can get them quite cheaply–ahem, so I've been told) and paint the tops of your keys in different colors–blue for the office, yellow for the house, etc. Then all you have to do is remember which color corresponds to which lock. If your eyes are incapable of differentiating between colors, why not try different patterns instead.

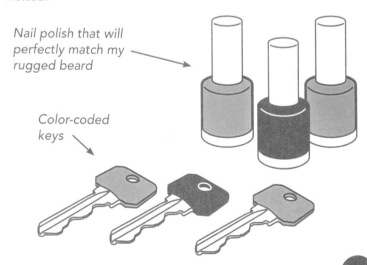

Nail polish that will perfectly match my rugged beard

Color-coded keys

STICKY TAPE PLACEHOLDER

Ever lose the place where the sticky tape starts? And once you've found it, do you struggle to get it unstuck? Annoying as hell, isn't it? Read on for the solution.

Before you finish with the sticky tape, get yourself a nail or pin and place it under the end of the tape. Now, when you want to find the end, it's clearly marked, saving you frustration and fingernails. Simple, but effective!

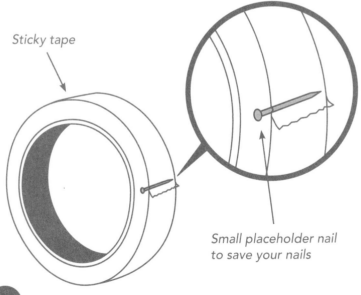

Sticky tape

Small placeholder nail to save your nails

STICKY NOTE CLEANER

Keyboards get all sorts of gross things stuck in between the keys–nose hair, food crumbs, boogers, etc. To avoid being the most disgusting person in the office, your keyboard needs to be cleaned regularly, and this hack shows you how.

Most offices have an abundance of sticky notes, which double as keyboard cleaners! Run the sticky side in between the keys and the glue with pick up the unhygienic detritus, saving you the embarrassment of a filthy keyboard and doing your bit to keep office germs down to a minimum.

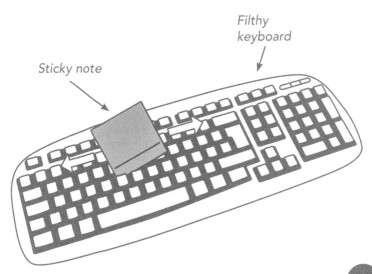

Filthy keyboard

Sticky note

PAPER CD CASE

What do you do when you need to send a CD to a client and the case has vanished? Use this hack! Not only will it enable you to transport your CDs safely, but you will never need to buy CD cases again!

Place a letter-sized piece of paper on your desk lengthways and position your CD at the bottom (in the center), half on, half off the paper–be careful not to scratch it at this point!

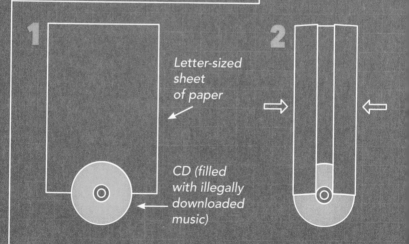

1

Letter-sized sheet of paper

CD (filled with illegally downloaded music)

2

Fold the left and right edges of the sheet in, around the CD, then fold the CD up and over, creating a little pocket. Now fold the larger flap down, over the CD. Unfold this flap and turn in its corners so you can insert the flap properly into the pocket you've made. There you have it, a perfect little CD wallet. You can even label it—just don't press too hard when you do!

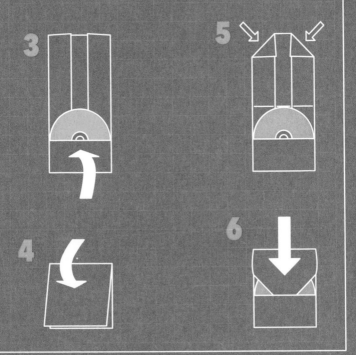

PLUG SMART

We've all done it. Disappeared under the desk to plug in yet another electronic device, which is supposed to make our working day go smoother, only to discover a baffling array of plugs. Which one do you remove? Do you randomly pull one out and hope the other end isn't attached to something important? What you should have done is this…

When you install a new appliance, carefully attach a bit of masking tape to the cord near the plug, leaving enough of an overhang so that it looks like a mini flag. Then write the name of the new bit of tech on the "flag" and you will always know which one is which. That's smart. That's plug smart. Got that?

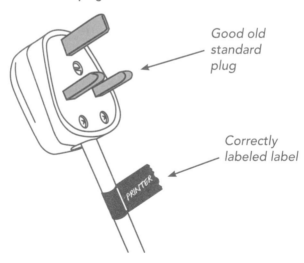

Good old
standard
plug

Correctly
labeled label

PRINTER

SPAM-AWAY

Spam. Not only a delicious canned meat that can be turned into fritters but also a source of constant annoyance in the form of electronic mail. If you have trouble with persistent junk e-mails, this will help.

To reduce the spam in your inbox, simply search for the word "unsubscribe." Among the gibberish and offers for discounts on bulk quantities of blue-pill substitutes, many junk e-mails include the option to unsubscribe, which, when clicked, will result in your being removed from the spammer's list.

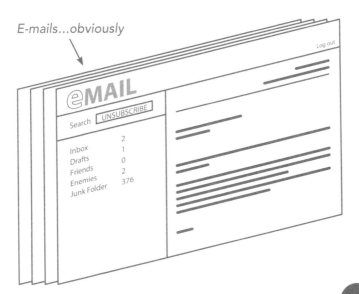

E-mails...obviously

PHONE HACKS

This chapter has nothing to do with phone-hacking scandals, so relax. Mobile phones have taken the world by storm–everywhere you look people are glued to the damn things! The other day I saw a family sitting 'round a table ignoring each other, all playing with their phones–or perhaps they were twerking each other on Twitbook? Either way, if you're a phone addict then you're going to love this next section.

AD-FREE GAMING

Smartphones have created thousands of new gamers, people who wouldn't ordinarily go near a console. This has changed the way advertisers work, and they now inundate us with pop-ups as we play our games. If this annoys you, you are not alone. Heck, my grandparents still mute the TV when the commercials come on!

To "mute" those damn in-game ads, simply switch your phone to airplane mode. (Unless, of course, you're online-gaming, in which case, your Russian bride is but a click away!)

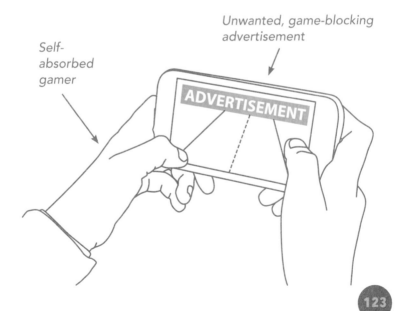

Unwanted, game-blocking advertisement

Self-absorbed gamer

ADVERTISEMENT

CHARGER HOLDALL

I love tripping over the mess created around the plug when my phone needs charging! Did the sarcasm show through there?

Keep your wires and phone out of the way with a clean plastic bottle. Remove the upper third of your chosen bottle (it should be big enough to house your phone and the charger wire) but keep the back of the bottle intact. Next, cut a hole in the back big enough to allow your charger plug to fit through. You now have a hanging plastic pouch for your phone.

Power supply

Non-hazardous charging phone

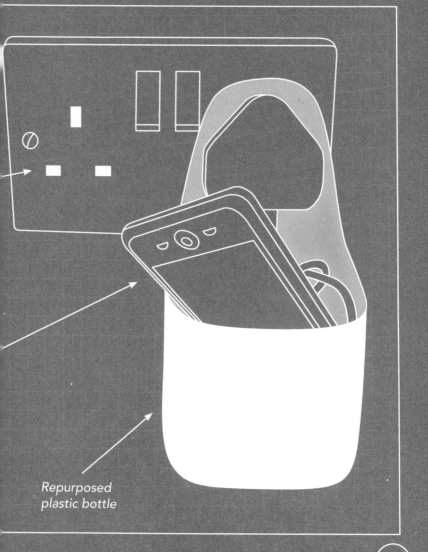

Repurposed plastic bottle

TOILET TUBE SPEAKER

Too cheap to buy speakers for your phone? Don't worry, I won't tell anyone. But I will tell you that a toilet paper roll works just as well. Alright, that's a bit of a stretch. But it does amplify the sound pretty well.

Cut a slot big enough to fit your phone and stop the roll from rolling away by sticking some pushpins into the tube to act as feet. Not only have you amplified the sound but you've also made a docking station. Aren't you clever!

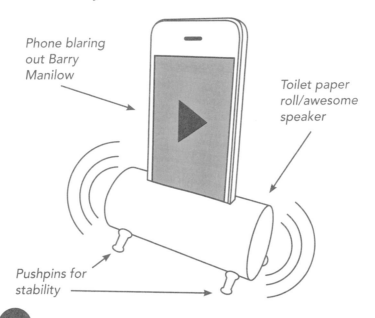

Phone blaring out Barry Manilow

Toilet paper roll/awesome speaker

Pushpins for stability

There are other cheapskate hacks to amplify your music. Why not try a glass...

...or even a bowl?

Both will work, but neither will look as amazing as your toilet tube docking station!

PHONE LANTERN

If you want to stay out after the sun goes down, you'll want to see what you are doing. Impress your fellow night owls by putting a water bottle on top of your phone to create a makeshift lantern. The light from the screen will turn the water bottle into a light so you can see in the dark while simultaneously attracting moths!

NB: Remember to turn off the lock on your phone to keep the light continuous, and make sure that the bottle has water in it for optimum refraction.

Water bottle

Moth-attracting light

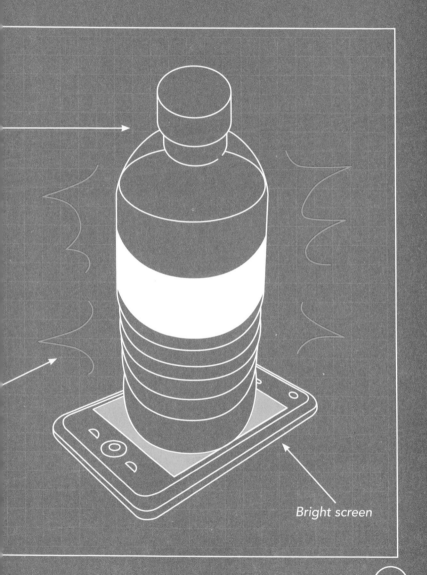

Bright screen

CHARGE BOOSTER

Just when you thought airplane mode couldn't get any more exciting, here's another hack that uses it. Turning on airplane mode while charging your phone will increase the speed at which your phone battery charges. Again, if you're waiting for a message from your new Russian bride, don't use this hack.

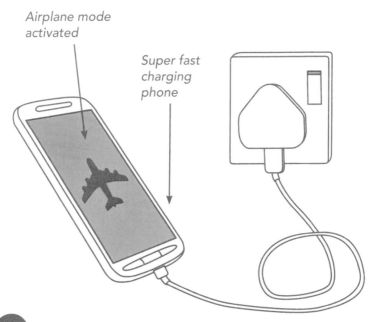

Airplane mode activated

Super fast charging phone

CABLE PROTECTOR

Nothing lasts forever, least of all a mobile phone charger. It is probably the most traveled of any charger, so will have seen its fair share of knocks. One common problem is when the wire frays, rendering it useless except as a component in a flux capacitor–and I don't see Doc Brown anywhere, do you?

To preserve your charger, wrap the spring from a ballpoint pen around the wire where it meets the connector; this will strengthen the wire and make it last longer!

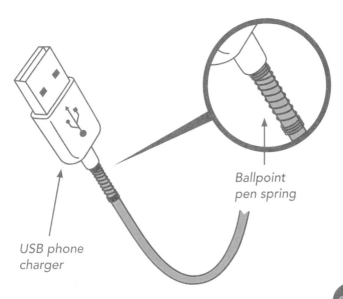

Ballpoint
pen spring

USB phone
charger

HACKS ON THE GO

From carrying your shopping with ease to not losing your precious place at the pub, these little life-hack gems will impress your friends and random members of the public alike. Kids on your street will begin shouting, "Hey, look! There goes MacGyver!" Well, they would, if they knew who MacGyver was…

THE COASTER
BEER SAVER

Even the most hardened pub-goers need to get up to pee now and again, at which point your beer and seat will be left unattended. Here's a sure way to safeguard both.

Simply place a coaster over your pint—this lets any overzealous glass collectors and would-be seat-stealers know that you've simply left the area temporarily. While you're up, you could even grab yourself a bag of tasty pork rinds! Mmm, pig skin.

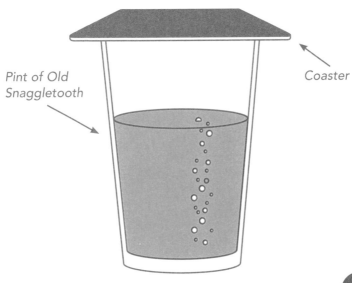

*Pint of Old
Snaggletooth*

Coaster

ANTI-THEFT NAP

Picture the scene: You're sitting on a train at the start of a long journey, so your best bet is to sleep through it. But how do you prevent someone unscrupulous from swiping your precious bag full of clothes and half-eaten sandwiches while you snooze?

Worry not. Simply put your foot through one of the straps, so if anyone does try to steal it you will be rudely awakened. Victim–1, Thief–0.

Bag full of half-eaten sandwiches

The leg of a person sleeping

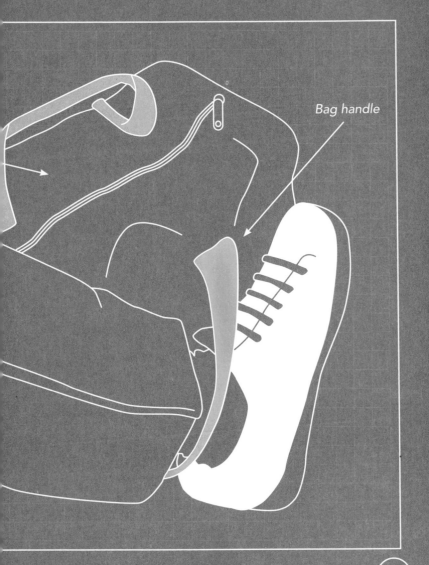

Bag handle

KARABINER BAG-HOLDER

"What's a karabiner?" you ask. Well, for all you nonclimbers, a karabiner is a roughly D-shaped metal ring that climbers attach their ropes to. You can pick them up fairly cheaply from outdoors stores and you'll need one for this hack!

When overladen with supermarket shopping bags, you can usually expect some kind of intense pain as the handles tear into the flesh on your fingers. Not anymore. Thread the handles of your bags onto your newly purchased karabiner and carry them all at once with no pain! Or you could just get your shopping delivered.

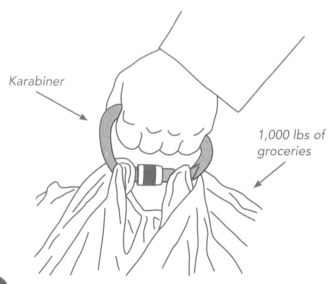

Karabiner

1,000 lbs of groceries

BONUS USB CHARGING POINT

Why do foreign countries insist on using oddly shaped plugs and sockets? Well, no matter–here's a hack to help all you gadget-loving vacationers.

If you're abroad and have forgotten the plug adapters, it's handy to know that most TVs nowadays have a USB port in the back, which you can use to charge your electrical devices. Unless you've gone camping.

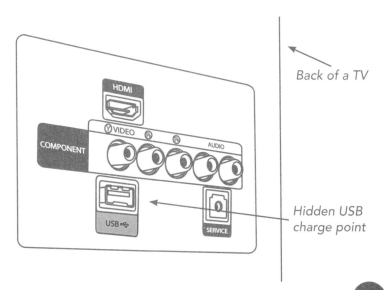

Back of a TV

Hidden USB charge point

STRAIGHT TO THE TOP

This hack is used by firefighters and police to get to where they're needed quicker–and you can use it too. If you want to use the elevator without stopping, press the "close doors" button until the doors are firmly shut and then, keeping it held, press the button for the desired floor. Keep the buttons pushed in until the elevator has started moving. Now stand back and relax as you travel uninterrupted until you reach your floor.

Elevator control panel

5 ◯ 6 ◯

3 ◯ 4 ◯

1 ◯ 2 ◯

B ◯ G ◯

◀▮▶ (DOOR OPEN) ▶▮◀ (DOOR CLOSE)

Press these buttons to go "straight to the top"

🔔 (ALARM) CAPACITY 11 PASSENGERS

EARPHONE IDENTIFIER

I can't stand it when earphone companies make their "L" and "R" symbols impossible to read–I have enough trouble getting them untangled! Here's a way to save time squinting at the backs of your earphones.

Simply tie a knot in the wire at the top near the earphone of your choice (make a mental note of which one, left or right!) and you now have a quick way to determine which side is which.

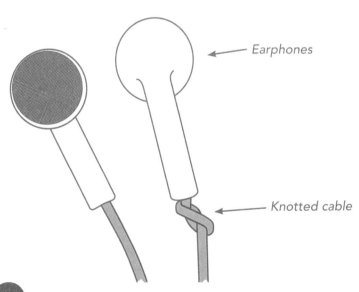

Earphones

Knotted cable

ELEVATOR FLOOR CANCELLATION

As a kid, my favorite prank was pressing all the buttons in the elevator before stepping out, leaving other poor so-and-sos to stop on every floor. This hack spoils such childhood fun, but it's probably something you should know.

If some young (or old) scallywag does press all the elevator buttons at once, all you have to do is press each button twice to cancel the floor stop. Take that, juvenile me!

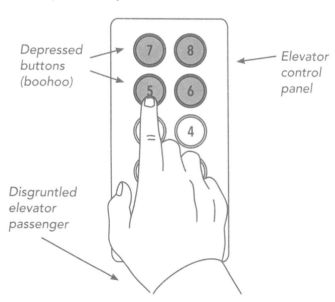

Depressed buttons (boohoo)

Elevator control panel

Disgruntled elevator passenger

THE NO-CUP-HOLDER CUP HOLDER

Your car hasn't got a cup holder–so what, right? I'm willing to bet, in reality, you would really like your car to have one for that essential morning coffee or long-journey refreshment. Well, here's how to get a free upgrade.

First, take your shoe off (not while driving, you maniac!). Now put it in the passenger footwell, wedge the toe end underneath the seat bottom and put your cup inside it. Yeah. Now you've got a cup holder. (Unless you wear six-inch stilettos to drive in.)

Hopefully not too stinky running shoe

Steaming hot cup of joe

LANGUAGE-BARRIER BEATER

If you're traveling abroad in a particularly exotic place with a complicated and unfamiliar language, it can be tough to even do the basics. Rather than spending four years mastering the different ways to ask for a "train station," use this hack.

Collect images of things you require–preferably of the item together with a description in the native language. So, if you want to order Tempura Dolphin in Tokyo, google it or take a photo of the item on a presourced menu and show it to the waiter; if you want to get back to that train station with a twelve-character name after a wild night in Amsterdam, take a photo of the platform sign, etc. The locals will know exactly what you mean and be so impressed that they might even give you what you ask for.

Foreign road sign

KETCHUP CAPACITY

We all know a few restaurants where they are too stingy to put bottles of ketchup, mustard, mayo, etc. on the tables. Instead they have sauce "dispensers" which make you squeeze the sauce into the smallest of paper cups with weird pleats in them. Invariably, you will have to go back for more halfway through your meal, and when you get back your mountain of fried food will be cold!

Did you know that the paper cups can be made bigger? Oh, yes indeed! If you unroll the top of the cup you can then spread out the sides and create a much larger area for your sauce. You even have enough space for two different types of sauce. Living the dream or what?!

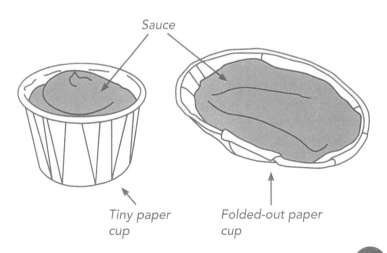

Sauce

Tiny paper cup

Folded-out paper cup

IN-CAR HOSTESS TROLLEY

This hack is for pizza lovers who can afford luxury extras in their car. If, like me, you live out in the sticks, takeout delivery isn't an option. How are you going to transport your pizza/burger/curry without it getting cold by the time you get home? Here's how.

Turn on the passenger-side seat warmer during the drive up to the takeout place (move all passengers to the back seat, obviously) and place your food on the seat for the drive home. When you finally get home your delicious meal will be piping hot. Let's just hope they haven't forgotten the extra jalapeños!

Button for seat heater

Piping hot pizza

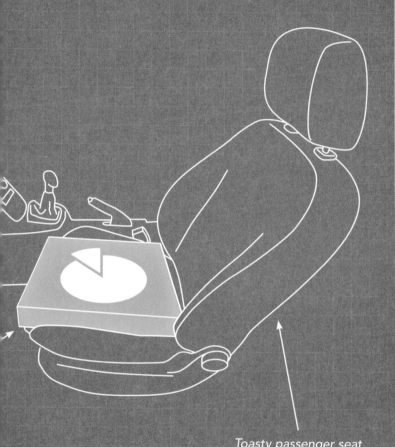

Toasty passenger seat

SECRET CASH STASH

I don't mean to scare you, but thieves are everywhere. There is probably one poised to steal this book out of your very hands!

To keep your cash safe when out and about, save your tube of lip balm once it runs out. You can then roll up your notes and keep them inside. Thieves are unlikely to want to steal lip balm and will probably just go for your wallet. Unless they've read this book of course, then you're in trouble.

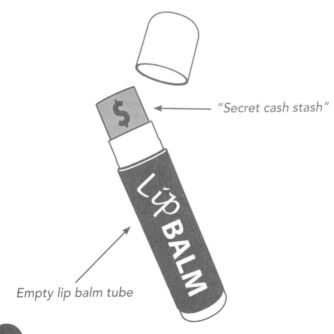

"Secret cash stash"

Empty lip balm tube

TANNING SECRETS

Here's another way of foiling those nasty thieves. It's a lovely day and you fancy lying out in the harmful rays of the sun, generally cooking yourself for a few hours. But what do you do with your valuables? Your Speedos don't have pockets!

Use an empty sun lotion bottle to store those precious things like car keys, your phone, or that picture of your pet Chihuahua, Foo-Foo. The light-fingered idiots won't even bat an eyelid as they walk straight past your valuables.

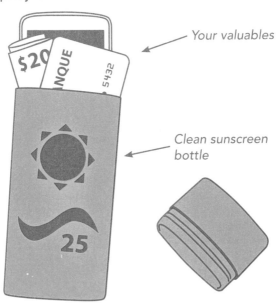

Your valuables

Clean sunscreen bottle

PLASTIC BAG BEATER

So, you're at the supermarket checkout and you're ready to start packing your groceries, only you can't seem to open a bag as the sides are stuck together. Why do they make them like this?! Well, no one wants to see you throw a tantrum in public, so here's how to achieve success.

Start by stretching the bag a little: Hold the top in the middle and pull on a handle. This separates the sides for you and you can proceed with your packing with no drama. I use this trick for garbage bags–they're rubbish! (Pun intended.)

Unwieldy plastic bag

HEALTH & WELL-BEING HACKS

Health is something we should take seriously. I try to go on six-mile runs as often as I can… OK, that's a lie, but I do play football. Although I can never remember which button is "shoot" and which is "pass!"

The hacks in this section are more about minor ailments than anything else, so if you just want some quick fixes for when you have a pimple or want to quit smoking then read on.

IBUPROFEN ZIT-AWAY

Tomorrow is the day of your important interview/wedding/party and, wouldn't you know it, an enormous spot has risen like the undead on your face. Think of the photos! The passive-aggressive comments from your so-called friends!

Worry not, turn to your friendly neighborhood ibuprofen. For this to work it has to be the liquid kind that comes in capsules. Stick a pin in one of the capsules to get to the good stuff and then liberally smother the liquid over your mini volcano. That zit will be gone in no time. (NB: Always read the label before using, and don't stab yourself with the pin, you dummy.)

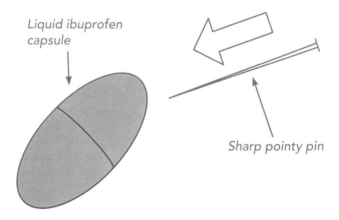

Liquid ibuprofen capsule

Sharp pointy pin

SOOTHE THAT BRUISE

Bruises are so unsightly and can be interpreted in all sorts of unsavory ways. Here's how to save yourself the embarrassment of having to explain them.

Soak a cotton ball in vinegar and place it on the offending area for sixty minutes or an hour, whichever comes first. After a while the bruise will disappear. Warning: I wouldn't recommend this method for a black eye–it'll sting!

Go on, guess what it is!

MALT Vinegar

Vinegar-soaked cotton ball

ZIP IT!

"Are you in the air force? 'Cause you're flying low!" This is something no one wants to hear shouted at them from across the street. If you are a fan of cheap trousers–or any other item of budget clothing that uses a zipper–then zip-slip is probably a daily occurrence. To keep your zipper up where it should be, take an old key ring and attach it to the tip of the zipper pull. When the zipper is up, slip the key ring over the button (or other loop or fastening) to hold it in place. Now innocent passersby will not discover the color of your underwear.

Key ring attached to zipper

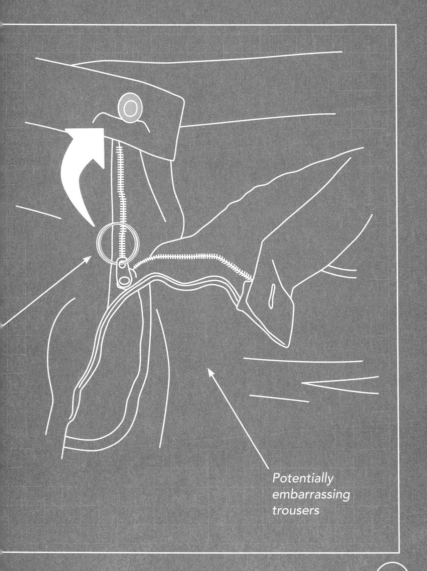

Potentially embarrassing trousers

SMALL STOMACH– SMALLER PLATE

Are you carrying around that extra "holiday weight" from five years ago? Or just in need of a beach body? Then I have a solution.

Doughnuts. Just kidding. To slim down and shed those unwanted pounds you have to eat less. (Shock! Horror!) To ensure this happens at mealtimes, put all your regular dinner plates in storage and break out the side plates (they're the smaller ones usually reserved for bread in posh restaurants). You'll find that you can't fit as much food on the plate, thus reducing your intake. Have a little discipline and don't go back for seconds!

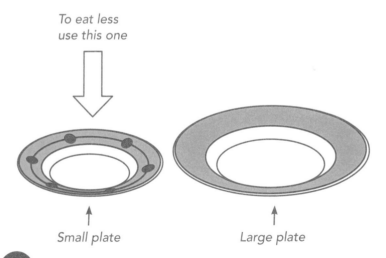

To eat less use this one

Small plate　　　　*Large plate*

ZESTY HEADACHE CURE

Stress is now a common factor in the workplace, leading to many a headache. To cure your pain you need a lime—no, not for your G & T!

Cut a segment of lime and rub it on your forehead. This will ease the pain, allowing you to soldier on like the martyr you are. A word to the wise, I would wait to be alone to try this. Getting caught rubbing fruit on your head might start some rumors!

Stress-relieving lime

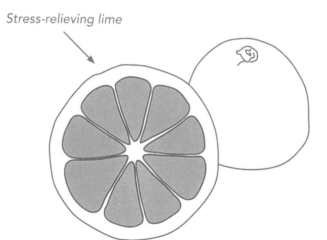

STOP SMOKING (SERIOUSLY, STOP IT)

As an established smoker I can say with pride that I've given up loads of times. It's the starting again I have a problem with. If you are serious about giving up then I recommend going to your doctor for some advice. This hack can't hurt though, just keep your towel on.

To give yourself a helping hand to quit, visit a sauna for three days in a row and you will sweat out the toxins, making it easier to quit.

Massive no-smoking sign just to drive the point home

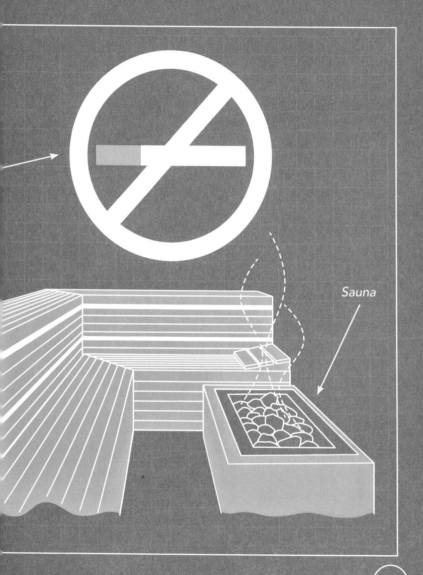

Sauna

SELF-REVITALIZER

All those long days at work (combined with late nights) can really take their toll. If you find yourself dozing off at your desk then use the following hack.

Sit up, open your eyes, and take a deep breath. Hold it in for as long as your lungs will let you and then breathe out, slowly. This increases your heart rate and sends oxygen to the brain, making you a bit more alert.

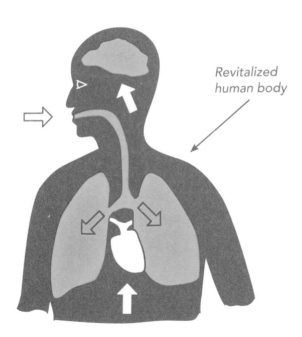

Revitalized human body

TEETH WHITENER

Do you want teeth white enough to blind random passersby? Then read on because this hack tastes as sweet as it sounds!

Mash up a strawberry with half a teaspoonful of baking soda until it forms a nice gooey paste. Now smear that sweet paste onto your teeth with your finger (make sure it's clean first, health and safety and all that). Spend the next five minutes trying not to lick it off. The malic acid in the strawberry will remove the stains on your teeth. Now rinse your mouth out (not with soap) and brush your teeth. Voila. Whiter teeth.

Juicy strawberry

Baking soda

BLESS YOU!

Not everybody likes sneezing, even if it is one-eighth the pleasure of an orgasm. If you're one of these people, try this hack.

To stop a sneeze, press your tongue to the roof of your mouth and it will vanish before you can say "a tissue."

A sneezer,
stifling a sneeze

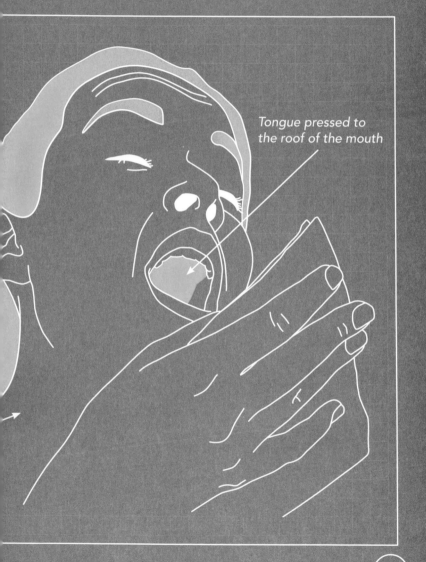

Tongue pressed to
the roof of the mouth

SORE THROAT SOOTHER

If you are suffering from a sore throat and your local shop has run out of lozenges, then get some marshmallows instead. They will do the trick as far as your throat goes. Just don't ask what they are made from, especially if you're vegan!

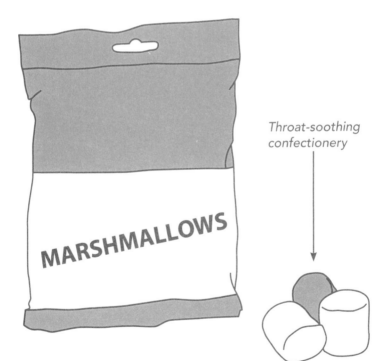

Throat-soothing confectionery

MARSHMALLOWS

NONDRIP ICE PACK

Sprained ankles and sore joints no longer have to be a soggy affair, but they will still be painful…sorry.

For a nondrip ice pack, soak a sponge in water then put it into a ziplock bag before freezing it. When you come to use it, the melting ice will collect in the bag instead of running all over your nice clothes and carpet.

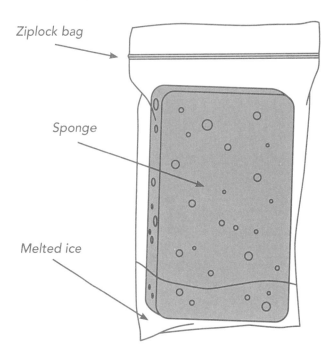

Ziplock bag

Sponge

Melted ice

ASSISTED SPLINTER REMOVAL

If you're one of those "outdoorsy" types (or a carpenter) then most probably at some point you'll get a splinter that won't come out no matter how hard you try. Here's a clever solution.

Pack a little baking soda into your travel first-aid kit (don't tell me you haven't got one!). When you do get that splinter, wet the area with water and sprinkle baking soda onto it, then cover with a plaster. Leave it for a day or so. When you peel back the bandage the splinter will be raised out of the skin, making it easy to pick out.

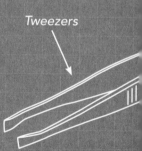

Tweezers

The offending splinter

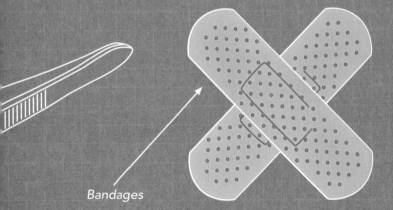

BAKING
SODA

Baking soda

Bandages

FRUITY COUGH SYRUP

Has anyone ever kept you awake all night with their infernal coughing? Have you ever kept *yourself* awake coughing? Either way, it's annoying as hell. Here's a way to help.

If you've had enough of coughing, don't just go and blow your hard-earned cash on the leading brand of syrup, simply drink some pineapple juice. It has a similar soothing effect–plus, it is one of your five-a-day!

Believe it or not, this is a pineapple...

...and this is its juice...

SWEET TONGUE SOOTHER

Some people (like me) simply can't wait to eat, even when something is clearly too hot to be scoffed. In goes the molten-hot food and bang goes your tongue, not to mention your dignity as you spit the food across the room.

Don't reach for the ice tray though, there's a much tastier way of soothing your pain. Reach for the sugar bowl instead. Put a teaspoonful on the burnt area and you'll feel a whole lot better. This won't cure your careless urges, these will be sticking with you for some time.

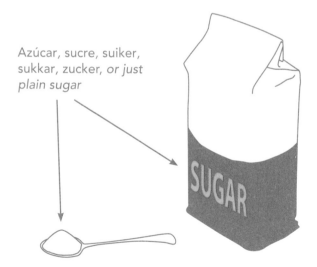

Azúcar, sucre, suiker, sukkar, zucker, *or just plain sugar*

GETTING AN EYEFUL

You've got something in your eye–an eyeball! If you have something else in there too it can be rather painful. Not good when you're driving the car. Pull over and get to a rest room.

Fill a basin full of water (make sure the basin is big enough for the next bit), and submerge your face. Open your eyes, and the foreign object will float out. Now you can get back to seeing and stuff. (NB: not recommended for contact-lens wearers who wish to continue seeing things.)

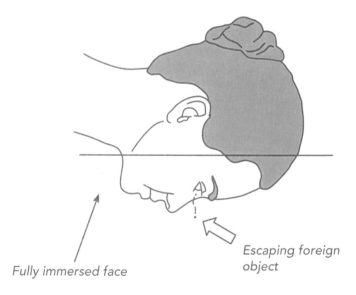

Escaping foreign object

Fully immersed face

RANDOM
HACKS

Here we have the more random hacks—an eclectic bunch, but that doesn't mean that they aren't good. Some of them are, admittedly, a little unlikely. But hey, I've given you some revolutionary stuff up until now—life-hack gold! So what's the harm in going a little off the beaten path?

CHEW YOURSELF CLEVER

Everyone needs a bit of help in exams but nobody likes a cheater. This hack will feel like cheating, but it isn't. It tastes good too.

When reviewing a particular subject, choose a particular flavor of gum to chew—my favorite is meatball. Chew the same flavor when you take the exam—this will improve your memory, as you associate the flavor with the subject.

When the exam is over, you can stick the gum under the table for the next candidate to use! I went through school without ever buying any gum thanks to the generosity of others.

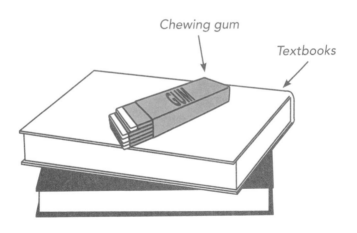

Chewing gum

Textbooks

FAKE NUMBER DETECTOR

If you suspect someone is giving you a fake phone number (despicable blackguards!) use this method.

Recite the number back to them incorrectly, i.e., change a digit. If they correct you then you know they are on the level. If they don't then you can chastise them for not having the conviction to just say no when someone asks for their number. Damn you, Mary Jane, damn yooooooooooou!

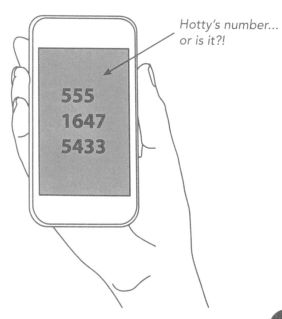

Hotty's number... or is it?!

555
1647
5433

RE-PING YOUR PONG

Who wants a game of ping-pong?! Well you can't, the ball is dented. It wasn't me, honest!

Removing the dents is actually pretty easy. Hold a lighter underneath the ball (not too close though), and the gases inside the ball will heat up and expand, forcing the ball back into its original shape.

If you don't have access to a lighter, putting the ball into a boiling pot of water will have the same effect. Just be careful when removing it, boiling water can be a little "burny."

Dented ping-pong ball

Lighter

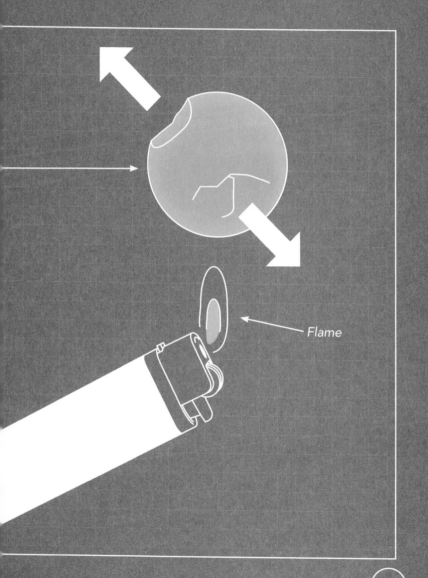

Flame

THE FRISBEE THROW

We've all seen athletic types parading about on the beach with a Frisbee, looking cool. But they can never throw the Frisbee quite right, can they? Here's your chance to beat them at their own game.

Get your perfectly untoned backside up off the sand and intercept that Frisbee. Or you could simply ask to join in, that would work too. To throw a Frisbee correctly remember to use the same action as you would when you whip a towel (come on, we've all done it!). The Frisbee will now fly as straight as an arrow. Congratulations, you are now part of their elite club.

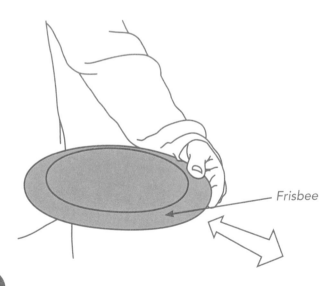

Frisbee

BORROWED TIME

Turning your house upside down to find that DVD you're desperate to watch won't help if you've forgotten you lent it to a friend—especially if it's the legendary *Goonies* (God bless you, John Matuszak).

If you want to remember what you have lent to whom, get your smartphone out and snap a photo or two. Get your friend to hold up the item they're borrowing as you take the photo and you will have irrefutable evidence of who's got what. Just remember to delete the photo when they have returned it!

Awesome movie

Forgetful Dave

CAR CRASH SAFETY SPLASH

A madman is driving straight at you with the intention of mowing you down. It could happen.

To avoid going under the car, where you will sustain life-threatening injuries, jump up before the car hits you. You will hopefully roll over the top of the car and your injuries will be less severe. If you're a little more athletic, you could try jumping over the car. I've seen it done but, admittedly, it was in a kung fu movie.

Speeding vehicle of a driver under the influence

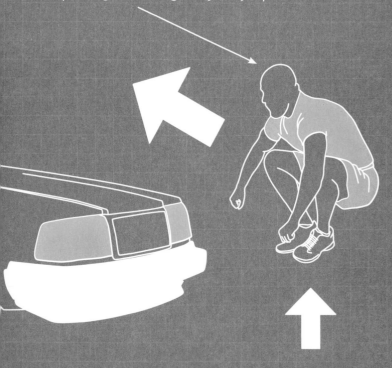

Unlucky/lucky car accident victim, depending on how high they can jump

TATTOO YOU DO

I've always wanted a tattoo, but I change my mind so much over what I want and where. I stick to transfers that my kids get with packets of sweets. They're less painful, and I can have a new one every week!

But if you want to design your own and have it for about a month, then follow these steps. Draw on your design with a permanent marker, rub baby powder over the area, and then spray on some hair spray. You will have your friends in awe. At the moment I am sporting a Mike Tyson-esque design on my face. I love it, but my boss is less keen.

Awesome
tattoo...
note to self,
"spell-check"

TALK TO THE ANIMALS

Zoo animals can be a little skittish, and I can say with authority that skunks don't like being handled. Luckily it sprayed the tourist next to me, which is a bonus. Not for her, though. If you want a better zoo experience, however, there is a way.

If you wear similar clothes to the zoo employees, the animals will trust you more readily and actually come up to you. I wouldn't try to get in with the tigers though; some lines aren't meant to be crossed. Especially lines with sharp teeth!

ZOO

Traditional zoo uniform, probably in khaki

BURIED ALIVE

This hack will help in the event of a natural disaster. What with global warming, crazy weather is happening all over the place—floods in your backyard, hurricanes in your mailbox, etc. Landslides, for instance, are occurring everywhere.

If you find yourself caught in just such a nightmare situation, and you have the time and your wits about you, take off your shirt and tie it around your mouth and face, this will protect your face from dirt, dust, stones, and gross bugs, thus preventing you from suffocating. Now all you have to do is dig yourself out like Uma Thurman in *Kill Bill*.

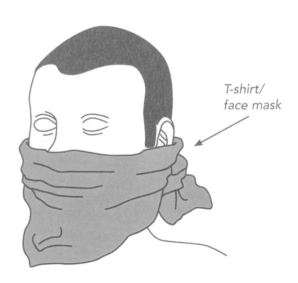

T-shirt/
face mask

TORTILLA CHIP KINDLING

The wilderness called and you answered. Your survival instincts so far have been great. You've built a shelter and caught some food— you've turned into a regular Bear Grylls. Well done, but how are you going to start the campfire without any kindling?

It's lucky that you remembered the tortilla chips. No, you're not going to eat them. Tortilla chips actually make great kindling. Pile them up and set them alight, you'll be eating in no time. Just not tortilla chips!

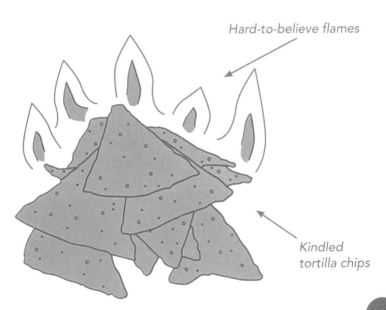

Hard-to-believe flames

Kindled tortilla chips

NO-CRAMP RUNNING

Being chased by baddies is bad enough, but what if you get cramp? They'll catch you, that's what!

Good guys never get cramp, probably from using this hack. If you exhale when your left foot hits the ground it stops you from cramping up. This is why bad guys never win; I'm sure of it!

Exhale

Left leg down

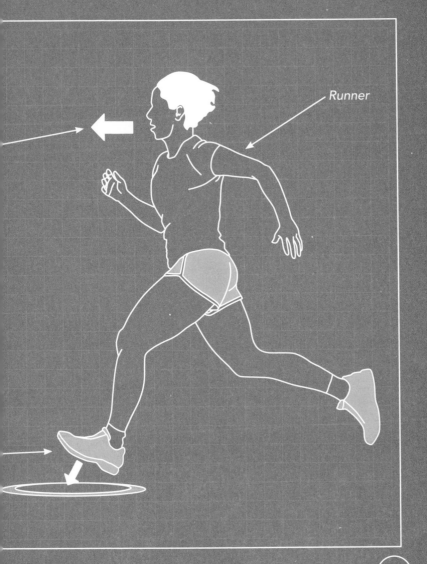

Runner

ALL TIED UP

You discover a damsel in distress tied up in a basement, it wouldn't sound good if you had to say, "Sorry baby, the knots are too tight and I can't undo them. I'll have to leave you here, but best of luck for the future."

If you really want to save her, you should twist the end of the rope to make it thinner. You will then be able to push the thinner rope through the knot, freeing the captive. She will be eternally grateful, and there might even be a reward!

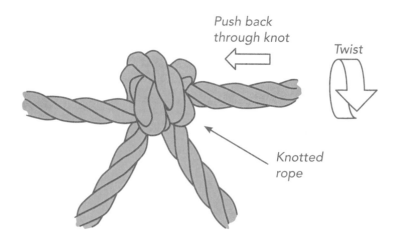

Push back through knot

Twist

Knotted rope

FINAL WORD

Congratulations—you are now a *Life Hacks* hero. Whatever life throws at you, you can handle it—whether it's a dripping paintbrush or a landslide.

Feel free to pass on these little nuggets of genius to all you meet and plugging the book wouldn't hurt. (I've still got to eat—I haven't found a hack for this yet.)

If you have some life hacks that are not featured in this book and think they deserve to be in print, e-mail them to auntie@summersdale.com.

Until next time—get hacking!

HACKS INDEX